Oryx Frontiers of Science Series

RECENT ADVANCES AND ISSUES IN BIOLOGY

Oryx Frontiers of Science Series

Recent Advances and Issues in Chemistry
Recent Advances and Issues in Physics
Recent Advances and Issues in Environmental Science
Recent Advances and Issues in Biology
Recent Advances and Issues in Computers
Recent Advances and Issues in Geology
Recent Advances and Issues in Meteorology
Recent Advances and Issues in Astronomy
Recent Advances and Issues: The Brain
Recent Advances and Issues in Oceanography

Each new volume in this series is the ideal first-stop source
for information on cutting-edge issues in the field.

Series features include

Recent research • Current social and ethical issues
New technology and applications • Biographies of key individuals
Documents, speeches, statements, and reports
Unsolved questions and research trends • Career opportunities
Organizations and associations • Print and nonprint resources
Glossary of terms • Comprehensive subject index

Oryx Frontiers of Science Series

RECENT ADVANCES AND ISSUES IN BIOLOGY

by Leslie A. Mertz

Oryx Press
2000

© 2000 by Leslie A. Mertz
Published by The Oryx Press
4041 North Central at Indian School Road
Phoenix, Arizona 85012-3397
http://www.oryxpress.com

Cover image of monarch butterfly caterpillar courtesy of Ben Czinski, photographer.

Library of Congress Cataloging-in-Publication Data

Mertz, Leslie A.
    Recent advances and issues in biology / by Leslie A. Mertz.
        p. cm.—(Oryx frontiers of science series)
    Includes bibliographical references and index.
    ISBN 1-57356-234-3 (alk. paper)
    1. Biology. I. Title. II. Series.

QH307.2.M47   2000
570—dc21   00-027080

CONTENTS

PREFACE

Over the last decade, the string of discoveries within the field of biological sciences has been staggering. Headlines trumpet cloning one day, the loss of biodiversity the next, and a new dinosaur species the day after. Genetically altered foods, the impact of global warming, the potential for life on other planets, and a host of other topics are vying for the world's attention. Biology—the study of life—has become a phenomenally dynamic scientific field that will have a profound impact on our path into the future.

Recent Advances and Issues in Biology offers an introduction to the field of biology and an overview of the research that has set the stage for some of the most important discussions of our time: How did life begin, and how did we get to where we are today? Why are some species disappearing? What can we do to ensure that Earth as an ecosystem survives for coming generations? Can gene therapy help people live better and longer? What avenues can we—and should we—pursue now that we have unraveled the genetic code of different organisms, including humans?

This up-to-date, easy-to-read, and engaging reference work pulls together information from hundreds of journal articles, popular-press reports, and political discussions to present a lively and concise portrait of the science. The vast scope of the biological sciences, coupled with the lightning-quick pace of the latest discoveries and research reports, made the selection of topics difficult. While *Recent Advances and Issues in Biology* does not cover every subject within this expansive field, it provides an opportunity for anyone with an interest in biology to see how the science has come so far so quickly, and what developments the future may hold.

The audience for this book spans new and veteran research biologists, biology professors and teachers, high school and college students, and members of the general public seeking clear and straightforward explanations of cutting-edge work in biology. *Recent Advances and Issues in Biology* also makes a wonderful companion to the textbooks used in introductory college courses and advanced high school classes, or as the sole text in courses that present a survey of the field as a whole. As a reference work, biology teachers will find it an excellent resource for injecting exciting new research findings and thought-provoking discussion points into their basic curriculum.

The book begins with five chapters that describe many of the important topics and research projects of the past few years. Comprehensive reference listings follow each section. Chapter 1, Biology Today, gives a sense of the field as a whole, research that has been done, and fresh courses of investigation. Chapter 2, Biodiversity, provides an overview of scientists' concerns about the planet, its species, and the relationship between the living and non-living worlds. Chapter 3, Ecosystem Management and Sustainable Development, explains how biologists and other scientists are developing a knowledge base that will help people make informed decisions about the environment of the present—and about the environment of years to come. Chapter 4, Evolution, describes how our view of the history of life on Earth is undergoing change as discoveries are made and novel technologies unfold. Chapter 5, Molecular Biology and Genetics, likewise contains information about exciting and promising technologies. Some of the most discussion-provoking topics, such as cloning and genetic engineering, are included.

Many of the subjects covered in Chapters 2–5 have generated heated debates in political, social, and economic circles, as well as among biologists and other scientists. Chapter 6, Social Issues, addresses some of these by summarizing the controversies along with the major political discussions and actions that have resulted.

A series of biographical sketches in Chapter 7 provides readers with more information about a number of the scientists whose work is featured throughout the book. The chapter covers both veteran researchers and those who are making important contributions in the earliest stages of their careers. Chapter 8 offers an insight into the biological sciences through a variety of key documents and letters. These materials are a reflection of scientific and public thought, and many have helped shape major policies influencing the field.

The next four chapters provide extensive resources for students, teachers, and scientists. Chapter 9, Career Information, gives budding biologists a view of the possibilities that await them, while Chapter 10 includes a collection of charts and other data that yield additional insight into the field. Dozens of organizations and many other information sources for new and veteran biologists are listed in Chapters 11 and 12.

The final chapter contains a glossary of terms, a number of which are either new or have only recently gained prominence among members of the scientific community.

Overall, *Recent Advances and Issues in Biology* gives readers a front-row ticket to the discoveries, the challenge, the promise, and the excitement that surround this timely—and timeless—field of scientific study.

CHAPTER ONE
Biology Today

B iology is firmly ingrained in the future. Cloning and genetic engineering have captured the world's attention. At the same time, almost-daily headlines tell of habitat destruction, the effects of pollution, or another species in danger of extinction. People everywhere are demanding cures for diseases. All these rely on the work of biologists. What challenges will they face in the next year, the next 5 years, or the next 50 years? What intriguing findings will they uncover? Which new questions will they raise?

Biology—the study of living things—is an expansive and rapidly changing field. What began hundreds of years ago primarily as observational study of plants and animals has evolved into a conglomeration of specialties. Biologists now investigate the most intricate workings of our bodies to learn what makes us tick. They conduct research into diseases and what they do to us, and decipher how bacteria and viruses live and reproduce. They have begun to unravel the genetic codes of many species, including humans, and are now beginning to manipulate those codes.

Biologists are still studying animals and plants in their natural habitats, and most are approaching their work with a renewed vigor. Their research continues to lead to discoveries of fascinating new species, but it has also found disturbing evidence that portends a less-diverse planet in the future. Some species are gone and others are on their way to extinction—and many times biologists have no clear understanding of the reasons for the demise. At the same time, field biologists have found other species thriving in environments that were once assumed to be harmful, if not lethal.

These field studies have led to a multitude of questions, and scientists in many fields, including biology, are taking a hard look at how Earth and its living things are impacted by human activity. One of the most complex questions is: How many people can Earth support? Biologists are dissecting ecological systems and the interlocking roles of animals and plants to understand how much change a specific ecological system can support before it collapses, and ultimately, human life is affected. They are working feverishly to predict the effects of human activity on the planet's biosphere and to recommend preventative measures to maintain a healthy environment for humans and other species.

Many insights into the future of life on Earth can be gleaned from the past, and biologists are exploring that realm, too. Evolutionary biologists are studying how life has changed over time, using a variety of techniques that have only recently come into existence. Their work has uncovered important information about why some species succeed and others fail. They have also identified several mass extinctions that nearly wiped out life on Earth. Perhaps that information will provide insight into the current loss of animals and plants, which many scientists believe signals another mass extinction.

Biologists are making stunning advances in genetic engineering and cell biology. Cloning has opened many possibilities, but it has also raised concerns and even fears about its misuse. Genetic engineering has presented unprecedented opportunities to fight human disease and to increase food productivity. Like cloning, genetic engineering is new, and its full ramifications have yet to be seen. Cell biologists are viewing this basic unit of life at a level of detail that was far beyond the imagination of scientists only a generation ago. As their comprehension of life's processes increases, they begin to understand the role of proteins and other chemicals in cellular processes. Some biologists use this knowledge to develop methods of interfering with diseases.

Other biologists consider each discovery as a page in the book of life, approaching every day in the laboratory with eager anticipation of a new finding. The charge of these scientists is filled with tremendous excitement and an unparalleled promise of new discovery, but accompanying this is enormous responsibility. In the next century, biologists will face great challenges. They will have to take into account the social, political, and moral framework of human society, and weigh how their work can best benefit the world and all of its inhabitants.

Excitement, promise, discovery, responsibility, challenge—all describe the current state of biology. Without doubt, that description will expand as swiftly as the field itself.

VALUE OF BIOLOGY

Descriptions of the many values of biological study are sprinkled throughout this book. An understanding of the relationship between species and a healthy environment, and the now-evolving field of biotechnology, has obvious benefits to human society. Studies of anatomy, physiology, genetics, and cell biology have provided great insights into the mechanisms and functions of the human body. Each year, physicians add many treatment options to their armory, and these advancements are prolonging human life.

In helping to determine the value of scientific studies, some scientists have begun to take on the daunting task of placing a monetary value on the very life systems at the heart of biology.

Biodiversity and Ecosystem Management

Scientists who wish to preserve global biodiversity are now seeing the advantages of stating, in dollars, what species and entire ecosystems are worth. Armed with such assessments, they are demonstrating the importance of biodiversity and ecosystem management to public policy decision-makers and to a populace that is sometimes swayed more by short-term gain than long-term planning. The figures are also helping to put species and ecosystems on the bargaining table as tangible and valuable assets.

Straightforward Benefits of Biodiversity

One of the most obvious and straightforward ways to calculate the value of biodiversity is to look at the income people directly derive from various species and ecosystems. Fishing is a good example. Fisheries employ 200 million people, produce 19 percent of all animal protein eaten by humans, and generate about $70 billion in revenues worldwide. When the world's fisheries experience declines, the ripples affect everyone from the fisherman to the consumer.

The forests have similar direct benefits. About one-third of Earth's land—some 3.5 billion hectares—is covered by forests. Before the rise of agriculture-based civilizations, forested land totaled about 5.5 billion hectares (1 hectare is the equivalent of about 2.5 acres). From those forests, 5 billion cubic meters of harvest goes to wood and fuel uses every year, and that amount jumps another 1.5 percent each year. Part of the harvest increase is directly related to human population levels. As the world's population grows, more land is needed for agriculture and other uses, and cleared forests are one source of that land. Like the fisheries, shifts in forest harvests can directly affect lumbering companies and employees, as well as the ultimate consumers of that forest product.

In both cases, sustainable development is crucial to the future of the industries. After all, what good is a forestry business without trees to cut, or a fishery without fish to catch?

Overall Value Estimates for Ecosystems

Several researchers have recently attempted to total the value of entire ecosystems. These estimates have received a good deal of attention in the past few years. Many scientists and environmentalists believe these figures are crucial in discussions of ecosystem management and habitat preservation. Others, however, point out that the estimates are based on extrapolations and determinations that can easily be challenged.

In 1997, ecological economist Robert Costanza of the University of Maryland, along with a team of other researchers, did their best to put a figure on 17 of Earth's ecosystems for 16 biomes. They derived their approximations from more than 100 previous studies of smaller areas and a number of original calculations, and they then extrapolated from those findings to determine the value of global ecosystem services. When they were unsure, they estimated on the low side. Nonetheless, the study placed the annual value of Earth's ecosystems at $16–$54 trillion. For comparison, they indicated that the annual gross national product for all the countries of the world is about $18 trillion.

The team that summed up the global values comprised a wide range of researchers, including representatives from the economics department at the University of Wyoming; the Center for Environmental and Climate Studies at Wageninengen Agricultural University in the Netherlands; the Graduate School of Public and International Affairs at the University of Pittsburgh in Pennsylvania; the geography department of the University of Illinois; the Institute of Ecosystem Studies in New York; the department of ecology, evolution, and behavior at the University of Minnesota; the environmental sciences division at Oak Ridge National Laboratory; the department of ecology at the University of Buenos Aires in Argentina; the Jet Propulsion Laboratory in California; the National Center for Geographic Information and Analysis at the University of California–Santa Barbara; and Ecological Economics Research and Applications Inc., in Maryland.

In the Costanza report, the researchers noted, "The economies of the Earth would grind to a halt without the services of ecological life-support systems, so in one sense their total value to the economy is infinite." Still, they felt that contributions were undervalued in policy decisions, because they weren't equated with "economic services and manufactured capital."

For their study, the researchers divided Earth's ecosystem services into 17 broad categories. The category "gas regulation" embodied the regulation of the chemical composition of the atmosphere, including the balance between carbon dioxide and oxygen, the benefit of ozone as a protection from ultraviolet-B radiation, and the levels of greenhouse gases. Another category, pollination, considered the benefits of pollinators in plant reproduction. For many of the valuations, the researchers considered how much the public would be willing to pay for the ecosystem service if it were available for purchase. They then multiplied the base value by the size of the ecosystem to arrive at a total. Although they acknowledged that their estimates were crude and likely too low, they felt their figures would give a better and much-needed indication of the worth of ecosystems.

Another report in 1997 estimated that the biodiversity in the United States was worth $300 billion every year. Led by ecologist David Pimentel of Cornell University, the research team assigned monetary values to ecosystem services, based on comparable human-provided services. For example, the team used the organic waste disposal costs of two cities— $0.04–$0.044 per kilogram—to give a conservative monetary value of $0.02 per kilogram of wastes recycled by decomposer organisms that do the same job. Multiplying that $0.02 per kilogram figure by the U.S. total of 3.1 billion tons of decomposer-recycled organic wastes, they arrived at a final value for the decomposers of $62 billion each year.

Using similar comparisons, the researchers included these estimates of annual benefits in the United States:

- soil bacteria and their role in generating topsoil on agricultural land—$5 billion
- nitrogen-fixing plants and bacteria—$8 billion
- natural means of pest control in forests—$5 billion
- animal pollination of agricultural and natural ecosystems—$40 billion

For the value of wild animals and ecotourism, they approximated the amount hunters, fishers, and other nonconsumptive recreationists, such as bird-watchers, spend on those sports, and assigned a value of $59 billion. When considering foods harvested from the natural world, Pimentel's group took into account commercial and sport fisheries, along with wild berries, nuts, maple syrup, and other foods, and estimated a total value of $3 billion per year to the U.S. economy.

In all, the research team reported that the value of the services provided by the biodiversity of the United States added up to approximately $319 billion a year, and the team noted that its estimate was likely low. The report concluded, "If future generations are to live in a safe, productive and healthy environment, sound policies and effective conservation programs must be implemented to protect biodiversity."

References

Botsford, L., et al., "The Management of Fisheries and Marine Ecosystems," *Science*, 25 July 1977: 509–15.

Costanza, R., et al., "The Value of the World's Ecosystem Services and Natural Capital," *Nature*, 15 May 1997: 253–60.

Dietz, T., and P. Stern, "Science, Values and Biodiversity," *BioScience*, June 1998: 441–44.

Mlot, C., "A Price Tag on the Planet's Ecosystems," *Science News*, 17 May 1997: 303.

Noble, I., and R. Dirzo, "Forests As Human-Dominated Ecosystems," *Science*, 25 July 1997: 522–25.

Pimentel, D., et al., "Economic and Environmental Benefits of Biodiversity," *BioScience*, December 1997: 747–57.

Zimmer, C., "The Value of a Free Lunch," *Discover*, January 1998: 104–05.

Introduced Species: Billion-Dollar Damage

Another way to place a value on Earth's ecosystem is by determining the costs incurred when humans purposely or inadvertently change it. While humans have deemed some introduced species, such as crops, as beneficial, other alien species have shown a darker side. Non-endemic species sometimes out-compete native animals and plants and lead to their decline and sometimes even extinction—Chapters 2 and 3 discuss several examples. The damage inflicted by introduced species, however, also extends to the world's collective pocketbook.

A study presented at the 1999 annual meeting of the American Association for the Advancement of Science indicated that harmful introduced species were costing the United States more than $122 billion a year. According to the study, conducted by Cornell's David Pimentel and three of his graduate students, introduced weeds resulted in annual losses of $35.5 billion in the United States. Non-indigenous insects added $35.5 billion in damages, including $22 million from the gypsy moth alone. Gypsy moth caterpillars attack a number of deciduous and sometimes coniferous trees, but they prefer a few forest varieties such as the white oak tree. The caterpillars inflict widespread defoliation in the forests of the northeastern United States. The study estimated other introduced species and the losses associated with them, including: European and

Purple loosestrife *(Lythrum salicaria)* is annually causing an estimated $45 million in damage to U.S. wetlands.

Photo by Leslie A. Mertz

Norway rats (*Rattus rattus* and *R. norvegicus*), $19 billion; Dutch elm disease, $100 million; and the marsh plant purple loosestrife (*Lythrum salicaria*), $45 million. The study also noted that about 42 percent of the species listed as threatened or endangered made the list after waging a losing battle with introduced species.

In a session on alien invasions at the same 1999 conference, a series of scientists described the harm done by introduced species and pointed to ballast waters from ships as a major vector for new animals. For example, Asian clams introduced into the San Francisco Bay swelled from 3 to 20,000 per square meter in a single year. The clams feed by filtering, and their number now allows them to filter all the water in the bay one to two times a day. Through this feeding method, they also concentrate toxic chemicals, one of which is poisoning the ducks and bottom-feeding fish that prey on the clams. Other ballast-transported invaders include the western Atlantic comb jelly, which was responsible for a crash in Black Sea fisheries, and the north Pacific sea star, which is hurting the shellfish industry in Australia.

During that meeting session, Thomas Fritts of the U.S. Geological Survey described the invasion of brown treesnakes, *Boiga irregularison*, in the Pacific islands. Native to New Guinea and Australia, the snake arrived in Guam after World War II. The island has since been virtually overrun. At one point, Guam had 10,000 snakes per square kilometer, he said, adding that the treesnakes had caused 1,658 power outages since 1978 and had bitten numerous residents. Fritts noted that 85 percent of the victims had been bitten in their homes while sleeping or watching television; a bite from the mildly venomous snake can cause sickness in children. Currently, the snake is spreading to other islands, where officials have launched eradication efforts such as traps, fumigants, and visual searches of arriving planes and ships.

Florida is affected by other species, according to session speaker Don C. Schmidts of the Florida Department of Environmental Protection. Florida's non-native species include more than 900 established, alien plant species; about 1,100 insects; 24 freshwater invertebrates; approximately 35 fish, three dozen amphibians and reptiles, feral pigs; and even two colonies of rhesus monkeys. He listed the effects of the introduced species as habitat destruction via such behaviors as rutting by the feral pigs, predation on native species, newly introduced diseases and parasites; genetic effects such as hybridization that may lead to species extirpation; and competition between the native and introduced plants and animals.

References

"Alien Invasions! Impacts and Control of Nonindigenous Species," session, American Association for the Advancement of Science 1999 annual meeting, Anaheim, CA, 24 January 1999.

Cornell University, "Alien Animals, Plants and Microbes Cost U.S. $123 Billion a Year," press release, 24 January 1999.

Evans, H., *Insect Biology: A Textbook of Entomology,"* Reading, MA: Addison-Wesley Publishing Co., 1984.

Pimentel, D., "The Impact of Non-Indigenous Species: An Illustration of Integrative Sciences," abstract, American Association for the Advancement of Science 1999 annual meeting, Anaheim, CA, 24 January 1999.

Biology's Non-economic Value

Although economic benefits are an important measure of the value of biological study, the field is valuable in many other ways. Results of research in this field can—and are—changing the world. Biologists are helping to maintain, and in some cases to restore and even to create, habitats for plants and animals. They are making groundbreaking findings that are affecting how healthy we are and how long we will live. The next four chapters provide a sense of the wide-ranging impact the biological sciences are having on the present and will have on the future.

CHAPTER TWO
Biodiversity

B iodiversity—a word that encompasses all life forms on Earth—has found a place in the language of environmental decision-makers. It enters debates on everything from forest management to pollution restrictions and from urban expansion to park development, where biologists have joined government officials, business leaders, and many others in collaborating on decisions that will have long-lasting effects on the Earth and on the organisms that live here.

Despite the overwhelming public belief that the greater the biodiversity, the healthier the overall environment, scientists have not been able to devise a master plan for protecting species. In fact, biologists still don't have a handle on such seemingly basic information as the number of species on the planet or the exact effect of a species' elimination from a particular ecosystem. The reasons for this lack of baseline data are many.

HOW MANY SPECIES?

The biggest hindrance to arriving at a total number of species living on Earth is that biologists know they haven't yet discovered them all. In fact, they aren't even close.

Field biologists are still locating new species of even large land animals, despite the extensive human invasion of the planet. Discoveries of smaller animals, like insects, continue at a rate so fast and furious that most receive little more than a name and notation of the location where they were found before being placed in storage in a museum. Likewise, the number of known plants is still rising. Because the documentation for many species is so minimal, and because those records are spread over such a large number of museums and other repositories, even the total number of *known* living plants and animals is just an estimate.

Currently, the number of known species on Earth hovers around 1.5–1.7 million. The estimated number of species yet to be identified ranges from about 3 million to 100 million or more. Sir Robert May, chief scientific adviser to the government of the United Kingdom and head of the U.K. Office of Science and Technology, prepared an estimate derived from an October 1997 meeting in Washington, D.C., called Nature and Human Society: The Quest for a Sustainable World. This estimate gives 7 million as the most likely total number of living species, with a "plausible range" of 5–15 million.

Of the 1.5–1.7 million known species of plants and animals, almost half are insects. All the other known animals combined total only about one-third of the number of known insects. A feeling for the number of insects, and beetles in particular, comes from Terry Erwin, an entomologist at the Smithsonian Institution. In Peru, he identified 1,700 different species of beetles in just one tree.

Higher plants, which include flowering plants and grasses, are also far down the list from insects, totaling about a quarter million known species. Most scientists agree that bacteria are likely a very diverse group; however, the number of known species is less than 5,000. Research led by Martin R. Fisk of Oregon State University in Corvallis estimates that the bacteria living below the Earth's surface, and particularly in rocks below the sea floor, may constitute as much as 50 percent of the planet's biomass. Likewise, marine biologists report that the oceans require a great deal more exploration to determine how many species they support. Known marine species total about 275,000, but estimates of undiscovered coral reef species alone run from 1 million to 9 million.

References

Fisk, M. R., et al., "Alteration of Oceanic Volcanic Glass: Textural Evidence of Microbial Activity," *Science,* 14 August 1998: 978–80.

Malakoff, D., "Seas Yield a Bounty of Species," *Science,* 25 July 1997: 486–88.

May, R., "The Dimensions of Life on Earth," in *Nature and Human Society: The Quest for a Sustainable World,* edited by Peter Raven and Tania Williams, Washington, DC: National Academy Press, 1999.

Revkin, A., "Why Care? While Scientists Grope for Practical Answers ..." *New York Times,* 2 June 1998.

Rosenblatt, R., "Into the Woods," *Time,* 14 December 1998: 60–65.

Tangley, L., "How Many Species Exist?" *National Wildlife,* December/January 1999: 32–33.

RECENT DISCOVERIES

Animals

In March 1999, science magazines touted the discovery in the Gulf of Maine of a deep-sea octopus with an unusual feature: bioluminescence. Named *Stauroteuthis syrtensis*, it periodically emits blue-green light from organs situated where suckers are normally located. In fact, co-discoverers Edie Widder and Sönke Johnsen of the Harbor Branch Oceanographic Institution in Florida, believe the light organs may have evolved from suckers. The light either blinks brightly, or it glows dimly and continuously for periods of up to five minutes. Scientists believe the bioluminescence, which is rare in octopods, assists in communication and in attracting prey.

In another part of the world, ichthyologist John Lundberg of the University of Arizona in Tucson set out in 1997 to study the Amazon River and located 35 new species of fish, including an electric fish that has a very peculiar diet: the tails of other electric fish. In all, Lundberg and his research team trawled the deep river for more than 2,000 miles and netted 375-plus species of fish.

Large mammals are also still being added to the list of known species. Since 1993, scientists have located three new species of muntjac, an antlered member of the deer family that lives in the forests of Vietnam. The three species, all part of the genus *Muntiacus*, are the truong son, the sao la, and the giant muntjacs. Plants discoveries are being made nearby, as well.

Plants

You might have guessed that—at least in North America—nearly all the plant species have been discovered. From 1975 to 1994, however, taxonomists averaged 60 new species of plants per year, and that's just for the portion of North America above the Mexico border, according to a report published by botanists Ronald Hartman and B. E. Nelson of the University of Wyoming.

Even large plants in conspicuous locations are making the list of new species. For example, in 1992, botanists Dean Taylor and Glenn Clifton "discovered" a large stand of shrubs next to a California highway. Taxonomist Barbara Ertter of the Jepson Herbarium at the University of California–Berkeley identified the plant, which was named Shasta snowwreath, or *Neviusia cliftonii*. It is part of a formerly single-species

Shasta snowwreath *(Neviusia cliftonii).*

Photo by Barbara Ertter

genus. Four years after the snowwreath discovery, Taylor found a new lily of the genus *Erythronium* just a few miles from Yosemite National Park.

Even the well-studied eastern United States has its share of unrecorded species. For example, a graduate student in 1998 found a new plant, *Fuscidea pallida*, while rock-climbing in a gorge near his university in North Carolina. The student ran only a dozen transects through the cliff ecosystem and found 23 genera, including the new species.

While most botanists agree that the heyday has passed for finding new species in Canada and the United States, both countries likely contain hundreds of undiscovered plants. Still, botanists and zoologists seeking new species have shifted much of their exploration to tropical nations, many of which have extensive areas largely unexplored by scientists.

References

Abrams, M., "Amazing Amazonians," *Discover,* January 1998: 88.

Johnsen, S., et al., "Light-Emitting Suckers in an Octopus," *Nature,* 11 March 1999: 113.

Krajick, K., "Scientists—and Climbers—Discover Cliff Ecosystems," *Science,* 12 March 1999: 1623–25.

Milius, S., "Unknown Plants under Our Noses: How Much Backyard Botany Remains to Be Discovered?" *Science News,* 2 January 1999: 8–10.
World Wide Fund for Nature, "Third New Mammal Discovered in Vietnam," <http://www.wwf-uk.org/news/news37.htm>, accessed 2 February 2000.

New Ecosystems

As biologists continue their search for new species, they are discovering bizarre forms of life in places previously overlooked or thought to be uninhabitable.

One unusual ecosystem is inside a cave in Romania, near the Black Sea. Described by a team of researchers, including biologist Thomas C. Kane of the University of Cincinnati, it holds four dozen animal species that rely for energy not directly or indirectly on the Sun and photosynthesis—the modus operandi for previously known terrestrial living systems—but on hydrogen sulfide. The bacteria and fungi that form the base of this extraordinary food chain make carbohydrates from hydrogen sulfide found in the cave. Kane and two other biologists from the University of Cincinnati reported the finding in 1996. Among the 48 animal species

Entering the Romanian cave that holds a newly discovered ecosystem.

Photo by Cristian Lascu; provided courtesy of the University of Cincinnati and Serban Sarbu

The large white dot (upper right) is a single cell of *Thiomargarita*, a giant sulfur bacterium. Above the cell is an empty part of the sheath, where the two neighboring cells have died.

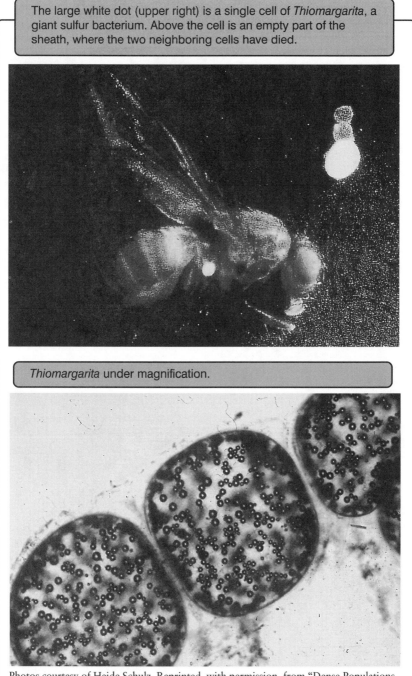

Thiomargarita under magnification.

Photos courtesy of Heide Schulz. Reprinted, with permission, from "Dense Populations of a Giant Sulfur Bacterium in Namibian Shelf Sediments" by Heide Schulz, et al., *Science*, 16 April 1999: 493–5. © 1999 American Association for the Advancement of Science they discovered were 33 new ones, including 4 spiders, a water scorpion, and a centipede.

In 1995 another group of scientists reported bacteria living in confined, underground aquifers filled with water that may be more than 35,000 years old. Todd O. Stevens and James P. McKinley of the Pacific Northwest Laboratory in Washington described the finding as an "active, anaerobic subsurface lithoautotrophic microbial ecosystem (SLiME) ... that appears to derive energy from geochemically produced hydrogen."

A report of a giant sulfur bacterium in sea sediment received attention in 1999. The scientists included discoverer Heide Schulz, a microbiologist at the Max Planck Institute for Marine Microbiology in Germany. The bacteria, *Thiomargarita,* was described as having spherical cells 0.75 mm in diameter, exceeding "by up to 100-fold the biovolume of the largest known prokaryotes."

References

Schulz, H., et al., "Dense Populations of a Giant Sulfur Bacterium in Namibian Shelf Sediments," *Science,* 16 April 1999: 493–95.

Skindrud, E., "Romanian Cave Contains Novel Ecosystem," *Science News,* 29 June 1996: 405.

Stevens, T., et al., "Lithoautotrophic Microbial Ecosystems in Deep Basalt Aquifers," *Science,* 20 October 1995: 450–55.

Wuethrich, B., "Giant Sulfur-Eating Microbe Found," *Science,* 16 April 1999: 415.

Archaea, a New Branch of Life

Unusual ecosystems, particularly the hot springs of the deep sea, spawned not only odd species, but an entirely new branch on the tree of life. The group of one-celled organisms that feed on hydrogen, sulfur, and salt are now grouped as *archaea* on the third branch. The other two branches are the *prokaryotes,* including the bacteria and blue-green algae (or cyanobacteria), which do not have a nucleus to encase their DNA, and *eukaryotes,* which are the nucleus-containing organisms, such as plants and animals.

The revelation that archaea warranted their own branch received validation in 1996 when scientists, including Carol Bult and J. Craig Venter of The Institute for Genomic Research in Maryland, described the genetic code of *Methanococcus jannaschii*, an archaea, down to the last of its 1.7 million bases, the chemical components of its DNA. The results not only contained enough never-before-seen genes to place the archaea in their own branch separate from bacteria, but it provided evidence that the archaea were actually more closely related to the eukaryotes than to the prokaryotes. The data suggested that the tree of life actually branches in two near its base, with the prokaryotes going one way and the eukaryotes and archaea going the other. Genetic data collected since then presents different versions of the tree, some with the bacteria and archaea on a branch separate from the eukaryotes.

References

Appenzeller, T., "Archae Tells All," *Discover,* January 1997: 37.

Chang, C., "Mapping a New-Found Life," *Audubon,* November–December 1996: 22–24.

Pennisi, E., "Is It Time to Uproot the Tree of Life?" *Science,* 21 May 1999: 1305–07.

Extraterrestrial Life?

In the past few years, the search for life has reached beyond our planet and into the solar system. Now known as astrobiologists, this small group of scientists are focusing particularly on Mars and on Europa, one of the moons of Jupiter.

ALH84001, Mars Rock

In 1996, several NASA scientists stunned the world when they announced that they had found what might be evidence of life on Mars. The evidence included microscopic, segmented, tubular structures, and mineral deposits that resembled the byproducts of Earth-based bacteria. All were found in a single small rock. The scientists explained that the rock originated on Mars, and that an asteroid collision had blasted the rock into space,

ALH84001, a rock collected in Antarctica that originated on Mars.

ALH84001,0

Courtesy of NASA

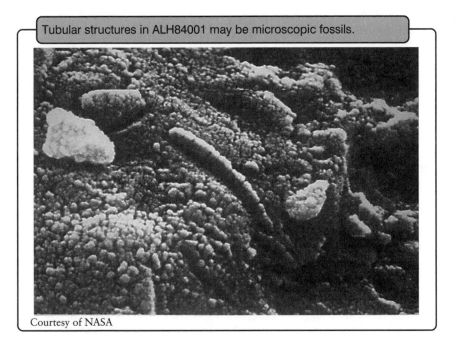

Tubular structures in ALH84001 may be microscopic fossils.

Courtesy of NASA

where it remained until roughly 13,000 years ago when it hurtled through Earth's atmosphere and landed in Antarctica. The research team, including David McKay of the Johnson Space Center in Houston, pointed out that the combined presence of several life-suggesting factors—each of which alone, it was admitted, could be attributed to other causes—had led to the conclusion that the rock contained evidence of life. Other scientists felt the confluence of the factors was merely coincidental and attributed each to strictly inorganic formation processes. A few months after the NASA declaration, a British team found similar evidence in another Martian rock.

Following numerous studies of the NASA rock, dubbed ALH84001, the scientific community has become increasingly skeptical that the structures and mineral deposits are signs of life.

Despite the doubts about ALH84001, some scientists are undeterred. In 1999, the same NASA scientists reported possible evidence of life on Mars in another meteorite, named Nakhla, which was discovered in Egypt in 1911.

Moons of Jupiter

Astrobiologists are now broadening their sights to the moons of Jupiter after the Galileo spacecraft sent back intriguing images of the moon Europa in 1996. The images appear to show an ocean with an ice crust. While the surface temperatures of Europa are much too low to sustain life

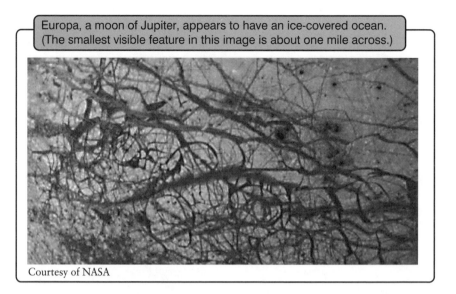

Europa, a moon of Jupiter, appears to have an ice-covered ocean. (The smallest visible feature in this image is about one mile across.)

Courtesy of NASA

as it is known on Earth, astrobiologists point to the suspected seas as cradles of life. Adding to the excitement is the 1999 discovery of the first planetary system orbiting a sun-like star. Located some 44 light years away, Upsilon Andromedae appears to have a three-planet system circling it.

Discoveries of functioning ecosystems on Earth in deep-sea vents and in other places where the existence of living organisms was completely unexpected are fueling hypotheses that life may not be unique to Earth.

NASA has even established a new Astrobiology Institute, which is managed by its Ames Research Center in Mountain View, California. Formed as a partnership between NASA and 11 other research organizations, the institute promotes and conducts interdisciplinary astrobiology studies and trains younger scientists in the field. (Information about the institute is available at its Web site, http://nai.arc.nasa.gov.)

References

Cowen, R., "Astronomers Find Planetary System," *Science News,* 17 April 1999: 244.

"Director of NASA Astrobiology Institute," advertisement, *Science,* 9 October 1998: 356.

Holden, C., ed., "Martian Life: Another Round," *Science,* 19 March 1999: 1841.

Kerr, R., "Requiem for Life on Mars? Support for Microbes Fades," *Science,* 20 November 1998: 1398–1400.

Marshall Space Flight Center, "Clues to Possible Life on Europa May Lie Buried in Antarctic Ice," <http://www.jpl.nasa.gov/galileo/news11.html>, accessed 2 February 2000.
Shreeve, J., "Find of the Century?" *Discover,* January 1997: 40–41.

IMPORTANCE OF BIODIVERSITY

The diversity of life is immense. This diversity matters for many reasons, including the all-important, and sometimes precarious, "balance of nature." In certain habitats, the loss of a single species can topple the elaborate equilibrium.

Keystone Species

When one species is key to balancing an entire ecosystem, that species is called a *keystone species*. A prime example is the sea otter.

The sea otters under study live along the coast of Alaska. These playful animals eat sea urchins. Sea urchins eat kelp. Many other species, both underwater and above, rely on kelp for survival. Led by marine ecologist James A. Estes of the Biological Resources Division of the U.S. Geological Survey and the University of California–Santa Cruz, a research team saw the effects of the loss of a keystone species when the number of sea otters in the Aleutian Islands plummeted by 90 percent during the 1990s, apparently because of predation by killer whales. Without the otters, the sea urchins decimated the kelp forests. The researchers speculated that the whales had begun hunting the otters after their usual prey, seals and sea lions, experienced their own decline, perhaps due to warmer water temperatures, overfishing, or other reasons.

The sea otter report gave a clear indication of the intricate web that connects living things. Normally, however, ecosystems do not offer such a revealing view of their workings. Biologists are left to surmise, estimate, and extrapolate using a wide range of variables, many of which hold their own uncertainties.

References
Kaiser, J., "Sea Otter Declines Blamed on Hungry Killers," *Science,* 16 October 1998: 390–91.
Stevens, W. K., "Search for Missing Sea Otters Turns Up a Few Surprises," *New York Times,* 5 January 1999: F1–F2.

New Medicines

Besides its importance for the balance of nature, biodiversity represents a nearly untapped resource for new medicines.

From Amphibians

Research just on amphibians provides many examples. For instance, the female gastric brooding frog, *Rheobatrachus silus*, swallows her eggs, shuts down her gastric system, and broods the eggs in her stomach. Information about how the frog accomplishes this feat could prove useful in the development of medicines to treat ulcers or other stomach disorders. Unfortunately, however, this particular frog is now believed to be extinct. The endangered Houston toad, *Bufo houstonensis*, produces alkaloids similar to the drug digitalis used by heart patients. The dendrobatids, or poison-dart frogs, secrete toxins that may be useful in research on nerve and muscle function. Likewise, the effects of ultraviolet radiation on frogs might shed light on the relationship between UV radiation and human health. Studies have shown that UV-B radiation has a detrimental effect on reproduction in some amphibians.

From Plants

Plants have been recognized for thousands of years as sources of medicines. Even the technology-driven Western medicine reaps the benefits: one quarter of all prescription drugs are derived from plants.

As habitat destruction continues—an area of rain forest the size of Yankee Stadium disappears each second—and plants are threatened with extinction, many scientists are scouring at-risk areas and interviewing indigenous healers about their plant knowledge. For example, one group of researchers has evaluated a type of tree bark and found three new drugs with anti-inflammatory properties. They hope to market one under the name "jungle bark" as a reminder to the medical community and the public of the importance of plants.

Additional values of biodiversity are discussed in Chapter 1.

References

Blaustein, A. R., et al., "UV Repair and Resistance to Solar UV-B in Amphibian Eggs: A Link to Population Declines?" *Proceedings of the National Academy of Science USA*, 1 March 1994: 1791–95.

Cox, P., "Conservation in Jurassic Park: Endangered Plants and the National Tropical Botanical Garden," topical lecture, American Association for the Advancement of Science 1999 annual meeting, Anaheim, CA, 22 January 1999.

Hallowell, C., "In Search of the Shamans' Vanishing Wisdom," *Time*, 14 December 1998: 71.

Sharp, B., "Frogs and Human Health," *Sanctuary*, March/April 1995: 8.

Tyler, M. J., "Declining Amphibian Populations—A Global Phenomenon? An Australian Perspective," *Alytes*, June 1991: 43–50.

Zug, G. R., *Herpetology: An Introductory Biology of Amphibians and Reptiles*, San Diego: Academic Press Inc., 1993: 183.

THREATS TO BIODIVERSITY

Considering the great benefits awaiting human discovery, as well as the pivotal role biodiversity plays in the balance of nature, scientists are more concerned than ever about the precipitous population declines among thousands of species and the extinction of many, many others. In some cases, a single species is dying off while related species are thriving. In other cases, whole animal groups are declining. One research team, led by Stuart L. Pimm of the University of Tennessee, calculated recent extinction rates to be "100–1,000 times their pre-human levels in well-known, but taxonomically diverse groups from widely different environments." That rate would increase tenfold if species now listed as "threatened" became extinct in the next century.

Scientists and environmental groups are working hard to intervene in population declines and species extinctions, in part by identifying species under threat, classifying their levels of endangerment, and detecting hot spots for biodiversity loss.

Reference

Pimm, S. L., et al., "The Future of Biodiversity," *Science*, 21 July 1995: 347–50.

CITES

The Convention on International Trade in Endangered Species of Wild Fauna and Flora, known as CITES, publishes two "appendixes" of various species to indicate the levels to which they are threatened with population declines and the extent to which they are regulated for international trade. Appendix I includes "all species threatened with extinction which are or may be affected by trade." Appendix II includes "all species which although not necessarily currently threatened with extinction may become so unless trade is subject to strict regulation; and other species which must be subject to regulation in order that trade in certain specimens of species referred to (previously) may be brought under effective control, i.e., species similar in appearance." In 1999, CITES listed a total of 821 species, 47 subspecies, and 22 populations under Appendix I; and 28,993 species, 100 subspecies, and 18 populations under Appendix II.

Population Losses

Other reports indicate similarly high numbers. In a report to the National Academy of Sciences in 1997, ecologists Jennifer B. Hughes, Gretchen C. Daily, and Paul R. Ehrlich of Stanford University reported alarming estimates of biodiversity loss in tropical-forest habitats. They began their analysis by calculating 220 as the average number of genetically or geographically distinct populations present in each species. Then, taking into account species density, the amount of forest lost each year, and other factors, they extrapolated to the staggering loss of about 1,800 populations every hour.

Another study in 1996 indicated distinct hot beds of potential extinction within the United States. Biologists, led by Andrew P. Dobson of Princeton University, concluded that 50 percent of U.S. plants and animals that are threatened with extinction occur in just 7 percent of the country. Hawaii, southern California, Florida, and southern Appalachia have particularly large numbers of threatened species.

A 1998 article totaled 110 extinctions in the United States "since humans started keeping track," plus another 416 suspected. It reported 469 animals and 675 plants listed under the Endangered Species Act. As an example of the magnitude of the problem, the article recounted that 120 of North America's 291 species of freshwater mussels are endangered, and about two dozen are extinct.

References

Adler, T., "Mapping Out Endangered Species' Hot Spots," *Science News,* 17 August 1996: 101.

Convention on International Trade in Endangered Species of Wild Fauna and Flora (CITES), "Protected Species" <http://www.cites.org/CITES/eng/append/species.shtml>, accessed 2 February 2000.

Hughes, J., et al., "Population Diversity: Its Extent and Extinction," *Science,* 24 October 1997: 689–92.

Mlot, C., "Population Diversity Crowds the Ark," *Science News,* 25 October 1997: 260.

Williams, T., "Back from the Brink," *Audubon,* November–December 1998: 70–76.

Amphibian Declines

Amphibians as a group have seen drastic losses since the 1970s. In addition to overall declines in populations, reports surfaced in the late 1990s about a high rate of deformities in frogs, including such malformations as additional legs and eyes, and incorrectly located eyes.

Bio-Indicators

Scientists are especially concerned because amphibians have long been considered bio-indicators of environmental conditions. They have been likened to the "canary in the mine," alluding to the age-old practice of miners bringing along a canary during their underground forays. If the canary died from otherwise-undetectable toxic fumes, the miners knew it was time to evacuate. Similarly, many believe that amphibian declines may signal dangerous environmental conditions for humans and other Earth inhabitants.

   Amphibians gained the title of bio-indicators for several reasons. They have an especially close relationship with their immediate environment. Additionally, the vast majority of them are biphasic; that is, they live part of their lives in an aquatic environment and part in a terrestrial environment. Many amphibians are herbivorous during one life stage and carnivorous during another. Their skin is permeable and is used to varying extents in respiration. Most species lay eggs that are permeable and unshelled, which provides exposure to the surrounding water or soil and to the Sun's rays. Changes in any part of the varied environments of amphibians (particularly wetlands) have the potential to wreak havoc on their populations. Since they have contact with so many aspects of the environment, they may also accumulate adverse effects from a number of environmental conditions.

Declines Suspected

Scientists first reported the widespread nature of the drop in salamander and frog populations during an international meeting of herpetologists in 1989. At this First International Herpetological Congress in Canterbury, England, field biologists began sharing stories of species disappearances or declines from habitats all over the world. Information about declines in the supposedly pristine environments of national parks and reserves added to the growing suspicion that these declines were worldwide in scope and had a global cause.

   A two-day workshop and a symposium in 1990 followed. Scientists presented anecdotal and empirical population-trend data to back up concerns about declining amphibian populations. They reported population dips from many areas, including eastern and western Canada, the far western and southeastern United States, the Rocky Mountains, Central America, the lowland Amazon Basin, the Andes, and Australia.

Declines Documented

Since that time, biologists have published the results of numerous studies to document declines and suspected extinctions. For example, a study of 27 sites in Sequoia and Kings Canyon National Parks in California from 1978–79 to 1989 revealed that the number of mountain yellow-legged frogs, *Rana muscosa*, dropped at all but one of the sites. Scientists studying two species of frogs in the Monteverde Cloud Forest Preserve in Costa Rica—the endemic golden toad, *Bufo periglenes*, and the harlequin frog, *Atelopus varius*—haven't seen any of them since 1990, and it is suspected that they are extinct. Similarly, scientists believe the southern day frog, *Taudactylus diurnus*, is extinct in its native Queensland, Australia. And the list goes on.

As reports of declines and extinctions have mounted, scientists have begun investigating the possible reasons. Suspected culprits range from increases in ultraviolet radiation to parasites, and from pollutants to habitat destruction. Most biologists now believe that different combinations of these factors are responsible.

References
Barinaga, M., "Where Have All the Froggies Gone?" *Science,* 2 March 1990: 1033–34.

Blaustein, A. R., and D. B. Wake, "Declining Amphibian Populations: A Global Phenomenon?" *Trends in Ecological Evolution,* July 1990: 203–04.

———, "The Puzzle of Declining Amphibian Populations," *Scientific American,* April 1995: 52–57.

Bradford, D. R., et al., "Population Declines of the Native Frog, *Rana muscosa,* in Sequoia and Kings Canyon National Parks, California," *Southwestern Naturalist* 39 (4), December 1994: 323–27.

Hayes, M. P., and M. R. Jennings, "Decline of Ranid Frog Species in Western North America: Are Bullfrogs *(Rana catesbeiana)* Responsible?" *Journal of Herpetology* 20 (4), 1986: 490–509.

Heyer, W. R., et al., *Measuring and Monitoring Biological Diversity: Standard Methods for Amphibians,* Smithsonian Institution Press, Washington, DC: 1994, 5–15.

Mlot, C., "Water Link to Frog Deformities Strengthened," *Science News,* 11 October 1997: 230.

Mossman, M. J., L. M. Hartman, R. Hay, J. Sauer, and B. Dhuey, "Monitoring Long-Term Trends in Wisconsin Frog and Toad Populations," in *The Status and Conservation of Midwestern Amphibians,* edited by M. J. Lanoo, Ames: University of Iowa Press, 1998, Chapter 21.

Pounds, H. A., and M. L. Crump, "Amphibian Declines and Climate Disturbance: The Case of the Golden Toad and the Harlequin Frog," *Conservation Biology,* March 1994: 72–85.

Richards, S. J., et al., "Declines in Populations of Australia's Endemic Tropical Rainforest Frogs," *Pacific Conservation Biology,* 1993: 66–77.

Wake, D. B., "Declining Amphibian Populations," *Science,* 23 August 1991: 860.

Wake, D. B., and H. J. Morowitz, "Declining Amphibian Populations—A Global Phenomenon? Findings and Recommendations," *Alytes,* June 1991: 33–42.

Wyman, R. L., "What's Happening to the Amphibians?" *Conservation Biology,* December 1990: 350–52.

Reasons for Declines

Scientists are attempting to protect against current and future species declines by determining the causes. Pollutants, introduced and invasive species, parasites, disease, habitat destruction, and overfishing are all possibilities.

Amphibians

Among the amphibians just discussed, causes for declines vary. For example, introduction of the predacious bullfrogs, *Rana catesbeiana,* and of game fish appear to have hurt the California red-legged frog, *Rana aurora*, and may have caused the extinction of the newt *Triturus alpestris lacusnigri*. In another study, scientists mimicked agricultural dosages of herbicides and pesticides to determine whether they could cause declines, and found that for the six species tested, those doses induced temporary paralysis. Other researchers tested a commonly used herbicide and pesticide cocktail on tadpoles from three U.S. species; all experienced paralysis or death. In Oregon, scientists found that a fungus, *Saprolegnia*, sickened amphibians. They hypothesized that one source of the fungus was hatchery-reared, introduced fish, which often are infected. Other studies found that acid rain could lower the pH of the water to lethal levels for amphibians.

A well-known study conducted in Oregon and published in the mid-1990s linked declines in the Cascades frog, *Rana cascadae,* and the boreal toad, *Bufo boreas*—both mountain species—with increased levels of ultraviolet-B radiation (UV-B). According to the study, led by Andrew R.

This Oregon Pacific treefrog *(Hyla regilla)* has an extra hind limb. Researchers believe trematode cysts caused the malformation.

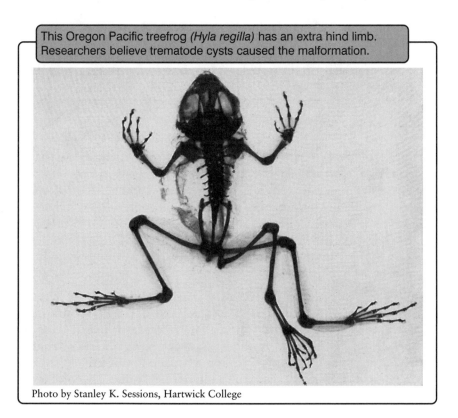

Photo by Stanley K. Sessions, Hartwick College

The trematode hypothesis for malformations in frogs was tested by treating tadpoles with trematode cysts. This multilimbed leopard frog (*Rana pipiens*) tadpole resulted.

Photo by Stanley K. Sessions, Hartwick College

Blaustein of Oregon State University, the UV-B decreased the survival rate of fertilized eggs by damaging their DNA. However, species from other parts of the country were unaffected by the heightened UV-B levels.

In 1999, amphibians again became front-page news when scientists found fungi and parasites as the causes of declines and malformations in specific populations. Taken together, two studies—one led by Stanley K. Sessions of Hartwick College in Oneonta, New York, and the other by Pieter T. J. Johnson of Stanford University in California—implicate infection by trematode parasites in the generation of multiple limbs in populations of five frog species.

Fish

Die-offs of non-amphibian species are also well documented and have suspected causes. Algal blooms in coastal waters have led to episodic declines in fish populations. In the Chesapeake Bay of North Carolina, the deaths of a billion crabs and fish were traced to toxins from *Pfiesteria*, an organism that reproduces into "blooms" or "tides." According to *Pfiesteria* co-discoverer JoAnn M. Burkholder of North Carolina State University, the organism has a complex life history and can change its tactics depending on the type of prey.

Land Snails

Another report attributed the death of the last known members of a species of land snail, *Partula turgida*, to an infection caused by a protozoan. Scientists were using a captive-breeding program to try save the South Pacific snail from extinction when the diminutive snail succumbed to the infection. With data collected by veterinary pathologist Andrew Cunningham of the Institute of Zoology in London and by parasitologist Peter Daszak of Kingston University in Kingston upon Thames, England, the report was the first to document an infection-caused extinction.

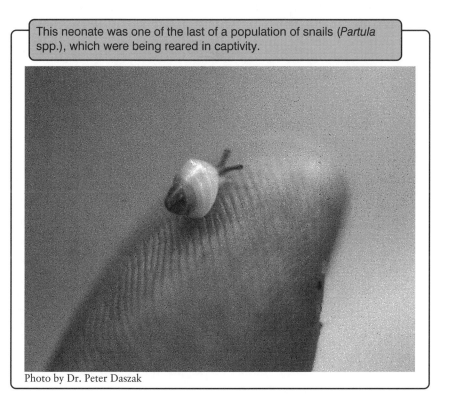

This neonate was one of the last of a population of snails (*Partula* spp.), which were being reared in captivity.

Photo by Dr. Peter Daszak

Songbirds

Ornithologists and amateur birders have a love-hate relationship—the emphasis is on the "hate"—with the brown-headed cowbird (*Molothrus ater*). Cowbirds are nest parasites, sneaking into the nests of much-smaller songbirds to lay eggs. The mother cowbird then flies off and leaves her egg in the care of the unsuspecting mother songbird. The cowbird's egg, usually larger than the foster mother's eggs, produces a similarly larger hatchling, which then competes with the smaller songbird hatchlings for food from the parent. The cowbird's size gives it the upper hand, and the songbird hatchlings often die of starvation in the care of their own parents.

Cowbirds have received increasing attention in the last decade because songbird populations are declining, and many observers place a good deal of the blame on cowbirds. One species at risk is the Kirtland's warbler, which has its summer home in a small area of central lower Michigan. In the last few years, scientists have begun taking a closer look at the cowbird and the behavior of the Kirtland's warbler foster parents to determine how they might be able to intervene against the cowbirds.

Two researchers studied whether cowbirds had a role in delaying the hatching time of the Kirtland's warbler eggs, which would give the cowbird an even greater advantage by being the first to hatch and begin growing. D. Glen McMaster of Saskatchewan Wetland Conservation Corp. and Spencer G. Sealy of the University of Manitoba found that the hatching time of the foster parent's eggs was indeed delayed by about 36 hours.

Ethan D. Clotfelter of the University of Wisconsin–Madison watched how female cowbirds go about selecting a foster parent's nest and leaving their eggs. For his study, he observed the female cowbirds parasitize the nests of red-winged blackbirds, although cowbirds parasitize the nests of more than 200 different bird species. He found that the cowbirds seemed to select blackbirds that made the highest number of calls when traveling to and from their nests, and he inferred that the cowbirds keyed on the blackbirds' vocalizations to locate potential nests.

In other work, scientists are trying to determine the distance cowbirds will travel to parasitize nests. Wildlife managers are using the information to try to set up cowbird-free zones around songbirds that will keep cowbirds away. Scientists Christopher B. Goguen and David Curson, also of the University of Wisconsin, have determined that the cowbirds will travel 8–10 kilometers to find a foster nest. Previous research had suggested 7 kilometers.

In addition, scientists are beginning to look at why some foster parent species take action against the intruding cowbird eggs by pushing them out of their nests or by piercing them with their beaks, but others do not. The answers to these and many other questions about the behavior of cowbirds and the foster parent species may help in the fight to protect the beleaguered songbirds.

Multiple Causes

The difficulties in determining the cause or causes of a decline become more evident when it is taken into consideration that each organism lives under a conglomeration of conditions. An individual can face exposure to a pesticide and a pollutant at the same time, or acidic water during one stage of its life and a pathogen at another stage. Two or three (or more) detrimental conditions can play off one another to weaken an animal or plant, enabling a final pathogen or fungus to deliver the fatal blow. For instance, ecologists at Cornell University in New York showed heightened global temperatures and pollution combined as a one-two punch against the sea fans *Gorgonia ventalina* and *G. flabellum*, corals found throughout the Caribbean. When the weakened corals subsequently faced a fungal disease, which alone would not have been fatal, they died. The report, presented by ecologist C. Drew Harvell at the 1999 meeting of the American Association for the Advancement of Science, reported 40 percent of these sea fan corals as either dead or at risk.

Other species migrate from one habitat to another during their lives, and the pressures they face in transit or in the different habitats can take a toll. In a study published in December 1998, Peter Marra and Richard Holmes from Dartmouth College in New Hampshire teamed up with Keith Hobson of the Canadian Wildlife Service in Saskatchewan to examine how one North American songbird is affected by the conditions at its wintering grounds. The three avian ecologists tracked American redstarts and the quality of food in their diets by analyzing their blood for levels of the isotope carbon-13. Low levels of the isotope correlate with a high-quality diet and the preferred habitat of wet mangrove or lowland forests. High levels signify a lower-quality diet and a dry scrub habitat. The data demonstrated that the male redstarts outcompeted the females 2:1 for the choice habitat. The effect played out in the weights of the birds. In the scrub habitat, the birds' weights decreased, while stress hormones increased. The weights of the birds that wintered in the wet habitats either stayed the same or rose. Additional examination indicated that the poorer the quality of the wintering grounds, the later those birds arrived for mating season the following spring, which diminished their breeding success.

Songbird research is helping to shed light on the plight of these particular birds, many populations of which are declining, and is also adding to the general pool of knowledge about the importance of habitat quality.

Natural Causes

Other threats to species diversity exist. The unpredictable nature of events like volcanoes and hurricanes rarely gives scientists the opportunity to measure before-and-after conditions. In 1996, however, biologist David A. Spiller of the University of California–Davis and a team of researchers had just completed a survey of lizard and spider populations on 19 islands of the Bahamas when Hurricane Lili struck. A day later, they repeated the survey; they also returned a year after the hurricane to study the species' progress. The islands that received the brunt of the storm lost species regardless of their population sizes, but on the islands that sustained only moderate hurricane forces, the larger populations fared better. In addition, the researchers found that the greater an animal's ability to disperse, the more successful it was at recolonizing. In other words, spiders recolonized faster than lizards.

The research team surmised that at one time in the past, all of the Bahaman islands, which were once connected, likely contained lizards. After the islands were separated by rising sea levels, hurricanes may have wiped out the lizard species on the smaller islands and caused the disparity between their numbers on the 19 islands. They reported, "Thus, given the poor dispersal abilities of lizards, their absence from most small islands may literally represent the high-water mark of previous hurricanes."

Biologists agree that Earth's species are under an assault of a grand scale. The next chapter will consider the role of biology in both local and global conservation efforts.

References

Blaustein, A. R., and D. B. Wake, "Declining Amphibian Populations: A Global Phenomenon?" *Trends in Ecological Evolution,* July 1990: 203–04.

Blaustein, A. R., et al., "UV Repair and Resistance to Solar UV-B in Amphibian Eggs: A Link to Population Declines?" *Proceedings of the National Academy of Science USA,* March 1994: 1791–95.

Burkholder, J., et al., "New 'Phantom' Dinoflagellate Is the Causative Agent of Major Estuarine Fish Kills," *Nature,* 30 July 1992: 407.

Carey, C., "Hypothesis Concerning the Causes of the Disappearance of Boreal Toads from the Mountains of Colorado," *Conservation Biology,* June 1994: 355–62.

"Coral Bleaching and Death Could Be Early Warning of Environmental Change, Cornell Ecologists Warn," <http://www.news.cornell.edu/releases/Jan99/AAAS.Harvell.hrs.html>, accessed 2 February 2000.

Corn, P. S., "Recent Trends in the Population of Wyoming Toads," *Froglog,* February 1993: 3.

Ferber D., "Bug Vanquishes Species," *Science,* 9 October 1998: 215.

Grant, K. P., and L. E. Licht, "Effects of Ultraviolet Radiation on Life-History Stages of Anurans from Ontario, Canada," *Canadian Journal of Zoology,* December 1995: 2292–2301.

Harte, J., and E. Hoffman, "Possible Effects of Acidic Deposition on a Rocky Mountain Population of the Tiger Salamander *Ambystoma tigrinum,*" *Conservation Biology,* June 1989: 149–58.

Hayes, M. P., and M. R. Jennings, "Decline of Ranid Frog Species in Western North America: Are Bullfrogs *(Rana catesbeiana)* Responsible?" *Journal of Herpetology* 20 (4), 1986: 490–509.

Johnson, P., et al., "The Effect of Trematode Infection on Amphibian Limb Development Survivorship," *Science,* 30 April 1999: 802–04.

Lannoo, M. J., et al., "An Altered Amphibian Assemblage: Dickinson County, Iowa, 70 Years after Frank Blanchard's Survey," *American Midland Naturalist,* April 1994: 311–19.

Laurance, W. F., "Is a Pathogen Decimating Australia's Rain Forest Frogs?" *Froglog,* June 1995: 3–4.

Milius, S., "Cowbirds Get Head Start with Egg Tricks," *Science News,* 28 February 1998: 135.

————, "Stealth, Lies and Cowbirds," *Science News,* 30 May 1998: 345.

Mlot, C., "The Rise in Toxic Tides," *Science News,* 27 September 1997: 202.

Pauli, B., "Environmental Contaminants and Amphibians in Canada," *Froglog,* February 1996: 2.

Raloff, J., "Overfishing Imperils Cod Production," *Science News,* 22 February 1997: 124.

Sessions, S. K., et al., "Morphological Clues from Multilegged Frogs: Are Retinoids to Blame?" *Science,* 30 April 1999: 800–02.

Smith, G., et al., "Response of Sea Fans to Infection with *Aspergillus* sp. (Fungi)," *Revista de Biologia Tropical,* 46 Supl., 1998: 205–08.

Spiller, D. A., et al., "Impact of a Catastrophic Hurricane on Island Populations," *Science,* 31 July 1998: 695–97

Wuethrich, B., "Songbirds Stressed in Winter Grounds," *Science,* 4 December 1998: 1791–93.

CHAPTER THREE
Ecosystem Management and
Sustainable Development

A s more is learned about life on Earth, biologists and other scientists are trying to determine how to maintain an environment that supports its seemingly endless variety. In particular, they are tackling tough questions about how humans and the planet's other species can coexist: How much can humans alter the environment and still keep the ecosystem relatively healthy? What is the most efficient use of the remaining natural areas? How many people can the Earth support?

In the past few years, concern about the future of life on the planet has risen sharply as reports of global warming, habitat loss, water and air pollution, species extinctions, and animal malformations have flooded the popular press and the scientific journals. *Sustainable development* has become the catch phrase for research into how to maintain an environment that supports the diversity of life, while acknowledging that human activities will continue to have impact. Angela Merkel, member of the German Parliament and Minister for the Environment, Nature Conservation and Nuclear Safety, provided this definition in an article she wrote for the 17 July 1998 issue of the journal *Science*: "Sustainable development seeks to reconcile environmental protection and development; it means nothing more than using resources no faster than they can regenerate themselves, and releasing pollutants to no greater extent than natural resources can assimilate them." The challenge, then, is to figure out how to balance the wants and needs of humans with the requirements of the Earth's overall ecosystem.

Reference
Merkel, A., "The Role of Science in Sustainable Development," *Science,* 17 July 1998: 336–37.

MANAGEMENT CHOICES

The vast assortment of life forms and the complexity of their interactions with other living things and with the physical environment combine to present an ever-changing set of conditions to the army of botanists, zoologists, and other scientists who are trying to manage Earth's ecosystems.

The number of species is enormous, and the variety is spectacular. The relationships among individuals of the same species, among different species, and among the ecosystem's living and non-living components (such as soil and water) are exceedingly complex.

What would happen if one part of that web of life were removed? Would the system collapse? Would the system continue to function, but on a less productive level? What if two parts were removed? What about three? What would happen if a new species were added to the web? If a wetland were drained or a forest logged, how much damage would occur? These are the types of questions that biologists ponder. Adding to the difficulty in answering them is the fact that every situation is unique: even two similar habitats located near one another have ever-so-slightly-different conditions: a few drops of extra rain one spring, a bit more runoff from an adjacent farm, a sunny spot where a large tree fell during a wind storm.

Biologists are investigating ecosystem management in many ways. Some are creating artificial, miniature ecosystems in the lab, in an effort to duplicate the natural environment, and then are introducing changes to see their effects. Others are taking an observational role in the field. Still other biologists are trying to control the changes in a natural environment or are running experiments in the field. Each approach has its advantages and disadvantages. Field studies are true to life, but they often involve many variables and may lead to incorrect conclusions; artificial studies, on the other hand, may limit variables, but they may omit key natural influences such as geochemical processes or other unknown factors. Both field and laboratory studies, however, can add vital information to biologists' comprehension of how ecosystems work.

References

Carpenter, S., "Ecosystem Experiments," *Science,* 21 July 1995: 324–27.
Lawton, J., "Ecological Experiments with Model Systems," *Science,* 21 July 1995: 328–31.
Roush, W., "When Rigor Meets Reality," *Science,* 21 July 1995: 313–15.

GIS

Conservation biologists consider Geographical Information Systems, or GIS, to be one of the most important technological additions to their field. GIS is technology that links a vast combined database geographically, or spatially, to Earth. If, for example, a biologist is trying to determine how best to protect an endangered animal, detailed data about the animal's specific habitats would be extremely useful information. Details might include anything from soil and vegetation types to the presence of rivers or the proximity of farm fields. Some GIS data are initially collected and located with GPS (global positioning system) technology, which is also becoming increasingly popular among field biologists. Using GIS, the biologist could select from already-collected, digital information about the area in question, overlay that data on a map, and then conduct an analysis of the best management methods. In other words, GIS brings together data from many sources and presents that information in a useful way. Researchers can select and combine the information that is pertinent to their studies.

Because scientists around the world are making their data available to other GIS users, the database is quickly expanding. The access to already-collected and digitized data, coupled with the array of software now available to manipulate that data, is giving researchers the opportunity to determine much more accurately the parameters surrounding and influencing their particular research projects.

Applications

The U.S. Fish and Wildlife Service (FWS) is one of a growing number of agencies and organizations that have embraced GIS as an important research tool. Like most other GIS users, they are turning to the system more and more as a provider of vital input for their studies. Some of the FWS initiatives that have used GIS to protect wildlife include the following:

- The use of GIS on a project in Hawaii helped to explain the habitat requirements of some of the state's 300 threatened or endangered species and enabled biologists to propose methods for their preservation. It also provided a view of historical population distributions, probable outcomes of management options, and a view of preservation objectives.

- The LaCrosse, Wisconsin, Fishery Resource Office applied GIS to create maps showing the movements of radio-implanted paddlefish (*Polyodon spathula*) from season to season. Paddlefish, long fish with paddle-shaped snouts, are important to the Upper Mississippi River System. Fishery managers can now refer to the maps when reviewing rehabilitation options and determining how they might be able to improve habitats for the fish.
- The Delaware Bay Estuary Project is using GIS to examine the distributions of vertebrates and of butterflies in a three-state area and their relationships to land cover and vegetation. The data are also helping project biologists to predict where species populations are located. The information gathered will help wildlife managers determine areas in need of protection and will enable land-use planners make ecologically sound decisions.

Challenges

GIS has a growing-circle effect: As more scientists use GIS in their projects, an increasing number will relay their findings back into the data pool. Provided with a growing database, more scientists will use GIS in their projects, and so on.

Perhaps the greatest challenge to GIS is the addition of new data, and the standardization of that data so that all GIS users can access it. Questions commonly revolve around how to employ a software program to relay collected data in an accurate, informative, and accessible format. Fortunately, GIS users can find and join electronic discussion groups, or they can turn to their colleagues who have volunteered to answer questions through various Web sites. The Wildlife Society, for example, has a Geographic Information Systems, Remote Sensing, and Telemetry Working Group to advance users' skills and increase their understanding of the technology. The U.S. Fish and Wildlife Service has a GIS Steering Committee and regional GIS coordinators. Users can also find information about GIS at numerous workshops and conferences held throughout the year.

National Biological Information Infrastructure

For extensive software tools to help run GIS, remote sensing, and other applications and processes, many biologists are turning to the National Biological Information Infrastructure, or NBII, an initiative of the Biological Resources Division of the U.S. Geological Survey. The NBII is becoming a main electronic storage facility for information gathered and then shared by public and private agencies, organizations, and institutions from within and outside the United States. Entry to the NBII's resources is available to anyone through its Web page at <http://www.nbs.gov/ nbii>. Although it was merely a glint in the eye of the National Academy of Sciences and National Research Council in 1993, the NBII is already becoming "the" information source for many users, including scientists, students, and decision-makers from the national level to local communities.

An example of the possibilities brought about through the NBII program is the joint venture between the U.S. Geological Survey and the National Park Service to create digital maps of the vegetation in some 250 national park units. Called the USGS/NPS Vegetation Mapping Program, the effort is generating a wide range of data that the NBII is making available to its users.

References

National Biological Information Infrastructure, "USGS/National Park Service Vegetation Mapping Program," <http://www.nbii.gov/factsheet/about/factsheet4.html>, accessed 2 February 2000.

U.S. Fish and Wildlife Service, "Current GIS and GPS Applications in the U.S. Fish and Wildlife Service," <http://www.fws.gov/data/gisapp.html>, accessed 2 February 2000.

U.S. Fish and Wildlife Service, "GIS Success Stories in the U.S. Fish and Wildlife Service, <http://www.fws.gov/data/success.html>, accessed 23 September 1999.

U.S. Geological Survey, "The National Biological Information Infrastructure," promotional booklet published by the National Biological Information Infrastructure, U.S. Geological Survey, U.S. Department of the Interior.

Wildlife Society, "The Geographic Information System, Remote Sensing, and Telemetry Working Group of The Wildlife Society, <http://fwie.fw.vt.edu/tws-gis/>, accessed 2 February 2000.

Species Insights

A starting point for many ecosystem research projects is acquiring a thorough understanding of the organisms involved, and scientists are learning startling new things every day. Here, observation is an important tool for gaining insights into animal behavior, diet, and species interactions.

At the 1999 meeting of the Society for Integrative and Comparative Biology, for instance, several biologists, including herpetologist James O'Reilly, announced that they had witnessed a suckling type of behavior among one of a group of wormlike amphibians known as caecilians. The young of this species, *Geotrypetes seraphini*, feed on either the mother's skin or skin secretions. Details about caecilian behavior have added importance because the numbers of these animals appear to be declining.

At the same meeting, physiological ecologist Harvey Lillywhite of the University of Florida reported that some snakes that use ambush as a tactic in prey capture can retain food in their digestive systems for long periods of time. The Gaboon viper, *Bitis gabonica*, held the record among the species studied: 420 days between defecations. The ecologist speculated that the added weight may give the snakes more leverage in battling large prey.

Knowledge about patterns of behavior is essential in the ever-growing conflict between native and introduced species. Hawaii is one of the hardest-hit areas: exotic weeds compete with native vegetation, introduced feral pigs tear up the ground, and non-native birds bring diseases to endemic species. A recent survey of the islands and their waters counted 21,368 species, nearly 20 percent of which are non-native—and that 20 percent figure does not include agricultural plants like sugar cane. Before biologists can begin to protect the native species, they must understand as much as possible about both the native and non-native species, particularly those aspects where the two overlap. That information may provide insights into how scientists can help a native species to compete better, or how scientists can control an invading species.

A study published in 1998 gave clues to the sometimes-perplexing success of species introduced into a new habitat. In this work, led by Ted Case of the University of California–San Diego, a team of researchers found that Argentine ant colonies relinquish some of their usual aggressiveness toward one another when they invade an unknown area. Instead of directing their energy to battles with one another, the formerly competing colonies focus on foraging and quickly grow in number. The researchers explained that this behavior compounds other positive benefits of the new habitat, such as a reduction in natural enemies and diminished competition from native species, and it helps the invading ants to thrive.

Governments are also recognizing the need for detailed knowledge about species. The U.S. government mounted fish surveys as part of the 1996 Magnuson-Stevens Fishery Conservation and Management Act. The act ordered regional fishery management councils to map essential fish habitat for more than 600 species and to identify threats to those habitats. To select essential habitats, scientists had to take into account nesting, feeding, and other behaviors. The act marked the first large effort by the U.S. government to incorporate habitat studies in fishery management.

In addition, government agencies are funding a myriad of research projects in hopes of curtailing the zebra mussel, a fingernail-sized bivalve that traveled from Asia to the Great Lakes of North America in the ballast water of a transoceanic ship. When the zebra mussel arrived in North America, it found an ideal habitat that was devoid of its natural predators. The series of events that grew from this accidental introduction has taken on catastrophic proportions. The mussels reproduced quickly, filtering food from the plankton-filled fresh water of the Great Lakes. The zebra mussels outcompeted the native mussels, which began to disappear. The microscopic plankton declined, the lake water became clearer, sunlight passed into previously murky depths, and seaweed began to flourish in the newly available light. Shallow waters became weed-choked and stagnant. Biologists across the United States and Canada are dissecting the zebra mussel's natural history to find a way to control its spread and to eliminate it from its newfound home.

References

Mertz, L., "Zebra Mussels: A Good Side?" *New Science,* Wayne State University, 1998: 12–16.

Mlot, C., "In Hawaii, Taking Inventory of a Biological Hot Spot," *Science,* 21 July 1995: 322–23.

Pennisi, E., "Meeting Spotlights Creatures Great and Small," *Science,* 29 January 1999: 623–25.

Schmidt, K., "Ecology's Catch of the Day," *Science,* 10 July 1998: 192–93.

Strauss, E., "Mutual Nonaggression Pact May Aid Ant Spread," *Science,* 30 October 1998: 854–55.

How Many People?

Another consideration in ecosystem management and sustainable development is the future size of the human race and how that size may modify the global environment.

Despite decreasing fertility rates the world over, Earth's human population is growing. Longer human life spans are adding to the number. Predictions vary, but many anticipate a 40 percent rise in the number of humans from 2000 to 2025. That means the population of 6 billion (in 2000) will bound to 8.5 billion (in 2025). The demand for additional space, food, and fuel will also increase. With such estimates, the global demand for food will double from the year 1990 to the year 2030, tripling in the poorest nations. A 1995 article by Joel E. Cohen of the Laboratory of Populations at Rockefeller University in New York reminds scientists and others that Earth's carrying capacity for humans cannot be determined with the same measures used for other living things, because humans make choices that can have great effects on the overall environment. Politics, demography, values, and economics can combine to raise or lower the Earth's carrying capacity.

As the twenty-first century dawns, another turning point has been reached: For the first time in history, more than half of all humans are living in urban areas. That percentage will likely increase, and developing countries will particularly feel its effects. In the United States, more than 75 percent are urban residents. This transformation indicates that urban areas are expanding and altering the character of formerly rural lands. It also means that the majority of humans no longer have close, daily relationships with the natural world, and they may come to have different attitudes about species and habitat preservation. The developers of future conservation efforts will have to take these changes into account.

With the growing human population, the roles of biologists in ecosystem management and sustainable development are many. Some biologists are developing or evaluating genetically engineered crops that may be more productive than current plants; others are studying the increased human contact with viruses and pathogens once associated with remote areas. The urbanization of tropical countries, for instance, brings increased threat from malaria-carrying mosquitoes. Biologists are also investigating development strategies that are less destructive to natural habitats, and they are studying the effects of pesticides and herbicides on many plants and animals, including humans. Their collective work is vast, and it is important. The research they are doing today will directly influence the world of tomorrow.

References

Cohen, J., "Population and Carrying Capacity: Beyond Malthus after Two Centuries," American Association for the Advancement of Science 1998 annual meeting, Philadelphia, 15 February 1998.

———, "Population Growth and Earth's Human Carrying Capacity," *Science*, 21 July 1995: 341–46.

Merkel, A., "The Role of Science in Sustainable Development," *Science*, 17 July 1998: 336–37.

Torrey, B., "Population Tectonics: Becoming an Urban World," American Association for the Advancement of Science 1998 annual meeting, Philadelphia, 15 February 1998.

EARTH'S ECOSYSTEMS

Marine Ecosystems

The world's fisheries have received mounting attention over the last decade as reports of population crashes became commonplace in the media and in scientific journals. At the 1998 meeting of the American Association for the Advancement of Science, Edgardo D. Gomez of the University of the Philippines presented an assessment of the world's fisheries. He reported that no undeveloped fisheries remained and that 60 percent of all fisheries were either mature or senescing. In addition, 69 percent of known fisheries' stocks were in need of immediate recovery work. Other reports listed half of the individual fish stocks as fully exploited, with nearly a quarter of them overexploited.

Fishing Bans

The solution is simpler in some cases than in others. In 1997, officials at the Florida Keys National Marine Sanctuary hoped to bring back its fisheries and those of surrounding areas by halting all fishing in the Western Sambos Ecological Reserve, a section of coral reefs and mangrove swamps. It marked the first time a fishing ban had been devised with the ultimate goal of replenishing the fisheries not only in the reserve but also in adjacent waters. Within two years, researchers noted improvements in the reserve, with increases in several species, including spiny lobsters and groupers, which are large, predatory fish. They are hoping that progress extends beyond the reserve's boundaries in the coming years. A similar ban in the Caribbean waters of St. Lucia led within two years to nearly double the fish biomass, compared to non-protected waters. Fishing bans in some other areas, however, have resulted in no increase in the quantity of fish. One ban, for instance, led to a decrease of more than one-third over a three-year period.

Fishing bans become even more complicated when proposed on a large scale because of the number of governmental jurisdictions and commercial fishers involved—and because scientific studies sometimes conflict.

A 1998 study of entire marine ecosystems, however, provided additional evidence in support of fishing bans. Daniel Pauly and Johanne Dalsgaard of the University of British Columbia in Vancouver, Canada, conducted the study with researchers from the International Center for Living Aquatic Resources Management in Makati, the Philippines. They analyzed fish catches among major fisheries worldwide over a 45-year period and found that humans had overfished the commercially valuable species, which are located at the highest trophic level (at the top of the food chain), and had begun shifting their focus to what were once considered "junk" fish. The new prey's location further down the food chain further weakens the ecosystem. From 1950 to 1994, the period studied, the researchers revealed a continuing decline of approximately 0.1 trophic levels per decade. The researchers concluded that the fisheries are not sustainable under current practices, and that fishing bans are imperative.

Trawling

Trawling is a case in point. This method of fishing, a technique that scrapes the sea floor for commercially valuable species like scallops, has come under increasing fire in the past decade. Scientists studying the effects of the practice say that 10–20 pounds of untargeted animals accompany every pound of commercial harvest. The untargeted animals include sponges, sea anemones, sea urchins, and other species that require long, undisturbed periods to recover. When trawling is repeated over the same area, however, recovery is curtailed. One report includes the endangered olive ridley sea turtle as a species affected by trawling. In India, researchers found that female turtles had to delay laying their eggs because trawlers blocked their path from the sea to their nesting beach. The result was that the turtles laid their eggs later in the year and in warmer-than-normal temperatures. The sex of turtles, like many other reptiles, is determined by egg temperature, and warmer eggs yield female hatchlings. Trawling, then, had an impact on the turtles by altering the hatchlings' female-to-male ratio.

Representatives of the fishing industry, on the other hand, point out that trawling is an age-old practice, and they refer to a study showing that although trawling has, in some places, eliminated such species as anemones, it has actually improved habitat for the highly valued Dover sole. Until additional scientific studies are available, government officials are left to struggle with uncertainties in their decisions on whether to ban trawling.

References

Botsford, L., et al., "The Management of Fisheries and Marine Ecosystems," *Science,* 25 July 1997: 509–15.

Gomez, E., "Ensuring the Sustainable Use of Marine Resources," session on "Science and Sustainable Development: Who Needs It?" American Association for the Advancement of Science 1998 annual meeting, Philadelphia, 15 February 1998.

Holden, C., ed., "The Ridleys Are Back," *Science,* 16 April 1999: 427.

Malakoff, D., "Papers Posit Grave Impact of Trawling," *Science,* 18 December 1998: 2168–69.

Raloff, J., "Fishing for Answers: Deep Trawls Leave Destruction in Their Wake—But for How Long?" *Science News,* 26 October 1996: 268.

Schmidt, K., "'No-Take' Zones Spark Fisheries Debate," *Science,* 25 July 1997: 489–91.

"The State of the Earth: 1995," *Discover,* January 1996: 80–81.

Williams, N., "Overfishing Disrupts Entire Ecosystems," *Science,* 6 February 1998: 809.

Freshwater Ecosystems

Fisheries

Freshwater fisheries have also seen declines. Up to one-third of the freshwater species native to the United States are threatened with extinction. Many of these are facing challenges from non-native and/or hatchery-raised fish that are added to waters to improve opportunities for sport anglers. One of the greatest dangers to America's wild, native fish is the fatal whirling disease, caused by a parasite believed to be spread by planted fish. The disease had spread through nearly two dozen states by 1998.

Hatcheries have, however, saved other species. The greenback cutthroat trout, listed as an endangered species in 1973, now has nearly two dozen self-sustaining populations in Colorado, thanks to a rearing program that brought it back from the edge of extinction. The trout is now the state fish, supplanting the introduced rainbow trout for the title.

Just as introducing fish can alter an ecosystem, changing other aspects of the environment also can have major effects. In 1963, for example, the Glen Canyon Dam was constructed on the Colorado River. A review of the river downstream from the dam showed that the dam not only decreased downstream river flow by 80 percent and sediment passage by 90 percent, but it eliminated the spring floods that annually cleansed the river of vegetation debris, churned up nutrients, and helped build beaches. On the recommendation of ecologists from the Bureau of Reclamation, which conducted the review, the dam was temporarily opened in 1996. The weeklong artificial flood rejuvenated the troubled ecosystem and created new habitats for some of its endangered birds and fish. Pleased with the results, ecologists began considering future floodings.

Wetlands

Another type of freshwater ecosystem that has been in the spotlight is the wetland. Humans are linked to the loss of 50 percent of Earth's wetlands, and more are endangered as people seek to develop additional land along lakes and rivers. The filling of wetlands—including the temporary springtime, or vernal, ponds—is often cited as a cause of amphibian declines. Biologists are also learning more about the critical role of wetlands as stopping points on bird migratory routes, as fertile bedding for a wide range of plants that support a diverse food web, and as water-purification systems. Estimates of wetlands' water-purification abilities place their value at nearly $3,000 per acre per year.

References

Egan, T., "Of Frankenstein Fish and the Call of the Wild: A Trout's Tale," *New York Times,* 2 June 1998: G10.

Mertz, L., "Survey of Amphibian and Reptilian Populations in Huron County, Michigan, with a Comparative Analysis of 1908 vs. 1996 Species' Richness and Relative Abundance" (dissertation), UMI Dissertation Services, Ann Arbor, MI, 1997.

Saunders, F., "Flooded at Last," *Discover,* January 1997: 64.

Williams, T., "Back from the Brink," *Audubon,* November–December 1998: 70–77.

Zimmer, C., "The Value of the Free Lunch," *Discover,* January 1998: 104–05.

Wetlands play a crucial role in the life cycles of many animals and plants.

Photo by Ben Czinski

Forest Ecosystems

Biologists are also providing scientific input to the heated debates over the use of forests. Like other ecosystems, forests are influenced by an enormous array of factors, both living and non-living. They are home to multitudes of plant and animal species, each of which exerts its own effects. Each forest is unique. A set of regulations for one forest may not fit another, even if that forest is similar and located nearby.

With such variables in play, biologists, forestry managers, government representatives, and others are wrestling with decisions about what is best for the forest and its species, while considering the role it plays for humans. As discussions continue, the forests are vanishing. One-third of the Earth's forests have disappeared since the rise of farming civilizations, and about 10 million hectares of additional land, mostly forests, are cleared each year to feed the growing world population. (A hectare is 10,000 square meters, or 2.471 acres.) Data from the United States are more positive, as many forests are recovering from the extensive logging of the early 1900s, and abandoned farm fields are reverting to woodlands.

Tropical Rain Forests

Tropical forests are particularly hard hit. These bastions of biodiversity are home to approximately two-thirds of all Earth's plant species. Since little virgin tropical rain forest will be saved from logging—the majority has already been logged or will be logged—scientists are either helping to save what is left, or they are conducting research to learn about the effects of logging methods and what changes, if any, are necessary.

A study published in 1998 provided evidence that, if left to regrow, some forests are able to regain most of their former variety. Conducted by Charles H. Cannon of Duke University and the Institute of Biodiversity and Environmental Conservation of the University of Malaysia, along with other researchers, the study contrasted undisturbed plots of rain forest land in Indonesian Borneo with plots that had been commercially logged eight years earlier. While the logged areas obviously had younger and therefore smaller trees, they held the same species diversity as the undisturbed plots. The researchers believe this example of the regenerative properties of rain forests should give hope to conservationists. Much additional research is necessary, however, to determine whether the same results will be seen in other tropical forest areas and following different logging methods.

Another long-term study reported in 1999 provided a closer view of the great diversity in tropical rain forests. A 13-year study of the Panamanian rain forest, conducted by Stephen P. Hubbell of Princeton University and colleagues from Princeton and the Smithsonian Tropical Research Institute in Panama, tested two popular hypotheses about diversity: intermediate disturbance and recruitment limitation. The former suggests that fallen trees create light gaps in the dense, closed-canopy forest, and those gaps provide the right conditions for many so-called pioneer species to grow. A series of other species later take the place of the pioneers as the gap fills in, finally leading to a mature forest. In summary, the intermediate disturbance hypothesis proposes that the biodiversity of the rain forest is fueled by the various stages of succession from pioneer species to those of the mature forest.

The recruitment limitation hypothesis, on the other hand, suggests that the inability of many rain forest trees to disperse over large areas allows a variety of trees to take hold. In such a situation, where success is measured not by one species outcompeting another for a resource, but by happenstance—being in the right place at the right time—a greater variety of species can survive. One tree cannot take over a habitat.

The Hubbell team examined 1,200 gaps in a 50-hectare plot of old-growth forest on Barro Colorado Island (BCI) in Panama. During the four complete censuses they conducted from 1982 to 1995, they "tagged, measured, mapped, and identified to the species level" more than 300,000 stems of 314 species. They also performed annual measurements of the gaps and the height of the canopy. While they found that new light gaps contained many seedlings, the overall diversity was identical to that of non-gap areas. They also found no difference in diversity as time passed and succession occurred. In addition, their analysis of individuals showed that each competed with, on average, 6.3 neighbors in their lives from sapling to canopy height. In other words, they competed with an average of only six other trees, or six species at most. The researchers concluded, "Thus, many or even a majority of the trees in the BCI forest canopy are likely to have won their sites by default." They acknowledged that the study considered only one tropical forest, but deemed their results were likely applicable to a broad range of tropical forests.

References

Cannon, C., et al., "Tree Species Diversity in Commercially Logged Bornean Rainforest," *Science,* 28 August 1998: 1366–68.

Chazdon, R., "Tropical Forests—Log 'Em or Leave 'Em?" *Science,* 28 August 1998: 1295–96.

Hubbell, S., et al., "Light-Gap Disturbances, Recruitment Limitation and Tree Diversity in a Neotropical Forest," *Science,* 22 January 1999: 554–57.

Moffat, A., "Temperate Forests Gain Ground," *Science,* 13 November 1998: 1253.

Noble, I., "Forests and Human-Dominated Ecosystems," *Science,* 25 July 1997: 522–25.

RESTORING DIVERSITY

Species Restorations

Reports of hundreds and hundreds of native species' extinctions and declines are tempered by success stories. Numerous species previously clinging to life are now rebounding, thanks to the efforts of biologists, other scientists, and a concerned public.

Like the previously discussed greenback cutthroat trout of Colorado that came back from the edge of extinction, the peregrine falcon once balanced precariously on the Endangered Species List. The falcon joined the list as a result of the pesticide DDT, which thinned its eggshells to the point that they broke beneath the brooding parents. DDT did the same to the eggs of other raptors. By the late-1990s, however, rearing-and-release programs brought the number of falcon breeding pairs back up to 1,600, a number considered sufficient for removal from the Endangered Species List. The success of the rearing programs are particularly evident in the eastern and the western United States. In the east, the birds had vanished by the 1960s. Three decades later, the area holds some 150 breeding pairs. In the west, the breeding programs brought the number of breeding pairs from 19 in the late 1970s to more than 800 in 1999.

Although sometimes successful, the rearing of animals or growing of plants can't magically wipe out extinctions. As described in the previous chapter, the attempt to restore the last known members of a species of land snail, *Partula turgida*, ended in the species' extinction when an infection wiped out the entire captive population. It became the first documented infection-caused extinction.

Given the large number of species that are disappearing or declining, coupled with limited resources available for restoring them, scientists have had to select those they feel are most "worth" saving. For that, they often turn to the population viability analysis, or PVA, developed in the 1970s to consider a wide range of factors in assessing the likelihood of success in attempting to restore a species or population. These factors include the species' probable fate, population dynamics, natural history information, and susceptibility to harming influences, among others.

While the PVA has been used extensively, many ecologists are now criticizing the analysis as inadequate. They describe some of the PVA's required data as difficult to obtain. Life-cycle data, for instance, is not available for some plants that remain dormant for years while awaiting a fire, flood, or some other irregularly occurring natural event to trigger growth. Others point out that the PVA doesn't consider genetic factors that can play a substantial role in smaller populations, such as those in captive-breeding programs.

In 1999, more than 300 scientists met to assess the PVA and consider options. Ecologists are now attempting to incorporate human influences like habitat destruction, population growth, and over-harvesting of species into the PVA.

Adding to the controversy, one ecologist at the meeting reported findings that some took to indicate that the PVA shouldn't be made more complicated after all. In this study, Gary E. Belovsky of Utah State University and his research team filled hundreds of containers with brine shrimp. Brine shrimp are of the genus *Artemia*, which has species found in salt lakes around the world. Different containers held different numbers of adults and varied food supplies. Then, the researchers waited. One by one, the shrimp populations died. By monitoring these mini-extinctions, the scientists were able to compare the results with what had been forecast by different PVA models. They found that the shrimp populations survived longer than the models predicted. They also learned that the carrying capacity of the container—the number of shrimp the container could support—was more important in determining the population's longevity than the number of adults initially placed in each container. The PVA, however, had placed more importance on the initial number of adults than on the carrying capacity. Finally, Belovsky reported that the simple PVA models matched the actual results more closely than the complex models. He and his research team are now planning to expand their experiments to more closely mimic natural conditions.

As the discussions continue, the population viability analysis will no doubt maintain its place as an important indicator in determining which species to target for protection. It will also likely receive some modifications.

References
Ferber D., "Bug Vanquishes Species," *Science*, 9 October 1998: 215.

Mann, C., and Plummer, M., "A Species' Fate, by the Numbers," *Science*, 2 April 1999: 36–37.

"Peregrine Falcon Recovery—Eastern & Central USA," The Peregrine Fund, <http://www.peregrinefund.org/conserv_epfalcon.html>, accessed 2 February 2000.

"Peregrine Falcon Recovery—Western USA," The Peregrine Fund, <http://www.peregrinefund.org/conserv_wpfalcon.html>, accessed 2 February 2000.

Williams, T., "Back from the Brink," *Audubon*, November–December 1998: 70–77.

Ecosystem Re-creations and Restorations

Despite the innumerable complexities within an ecosystem and the often perplexing uncertainties about how they work, some scientists are undertaking the enormous task of attempting to re-create them.

Pleistocene Park

One ecosystem re-creation with particularly grand expectations is Pleistocene Park, a project in which a group of Russian, American, and Canadian biologists and ecologists are attempting to turn back the clock. The scientific team includes Russian ecologist Sergei Zimov, Terry and Mimi Chapin of the University of Alaska–Fairbanks, and several other researchers. In a 160-square-kilometer area of Siberia, the international team hopes to recreate the mammoth steppe, an ecosystem common in Siberia during the last Ice Age, but unknown for the last 11,000 years. Their attempt involves introducing grazing animals to the area to tear up the plants of the current ecosystem and permit the re-establishment of grasses appropriate for a mammoth steppe ecosystem. A herd of Yakutian horses is already at the park. Bison, the last of which walked Siberia two millennia ago, will also arrive at the park, along with moose and reindeer.

Acknowledging that they cannot reproduce the cooler climate or the mammoths of the Ice Age, the team members still believe the project has merit. They speculate that the change in vegetation from moss to grass will lead to a return to a drier soil reminiscent of the Ice Age era. Once the bison are established at the park, the team will consider adding such predators as the now-threatened Siberian and Amur tigers.

Salton Sea

The Salton Sea of southern California poses much different questions. They run the gamut from "How should we save it?" to "Should we save it at all?" The Salton Sea is a large desert lake about halfway between San Diego and Arizona's western border. A remnant from an artificial rerouting of the Colorado River that occurred early in the twentieth century, the lake was a tourist destination for a short while. Unlike the life-giving waters characteristic of most lakes, the Salton Sea has become a witch's brew of deadly potions like hydrogen sulfide and ammonia submerged in water laden with a 25 percent higher salt concentration than the ocean. In the last decade, more than 200,000 birds at the lake have died from avian cholera, botulism, or other causes not yet identified.

Some scientists and politicians believe the lake should be restored. As one of the few wetlands left in southern California, it is important to wildlife. Birds rely on the Salton Sea as a stopping point during their long spring and fall migrations. Some politicians also favor the restoration of the lake as a boon for tourism. Other biologists, however, believe that restoration efforts are misplaced at the Salton Sea; instead, they recommend that the focus be shifted to other wetlands that would be safer havens for animals.

Despite the controversy, the U.S. Congress passed the Salton Sea Reclamation Act in 1998 to determine possible solutions to the problems that have developed at the desert lake. A team of scientists is collecting data and plans to make recommendations.

Aral Sea

Another ecological nightmare is rearing its head in Uzbekistan as the Aral Sea—once one of the largest lakes in the world—continues to shrink. Since 1960, the lake has lost 80 percent of its water to the irrigation of nearby cotton croplands, many of which carry high concentrations of DDT and other pesticides. An increase in salinity accompanied the decrease in water volume. The now-dried and salt-covered seabed is an agricultural deathbed. Once about the size of Lake Huron, the Aral Sea is likely to withdraw further and separate into three lakes early in the twenty-first century. Scientists are now trying to determine the best ways to preserve what's left of the sea, and they are considering diking and refilling part of the lake as a possible solution.

In both of these cases, the "right" answer is elusive. Biologists, however, are among the ranks of scientists who are adding to the pool of knowledge that will help cut the path for the future of the Earth's ecosystems.

References
Ellis, W., "Soviet Sea Lies Dying," *National Geographic,* February 1990: 73–93.
Kaiser, J., "Battle over a Dying Sea," *Science,* 2 April 1999: 28–30.
Stone, R., "A Bold Plan to Re-Create a Long-Lost Siberian Ecosystem," *Science,* 2 October 1998: 31–34.
———, "Coming to Grips with the Aral Sea's Grim Legacy," *Science,* 2 April 1999: 30–33.

GLOBAL WARMING

One of the most-discussed worldwide ecological phenomena is global warming. While report after report provides evidence that Earth's temperatures are rising, scientists are hotly debating the extent of that increase and its long-term effect on life.

Global warming occurs when so-called greenhouse gases, such as the carbon dioxide produced by the burning of fossil fuels, build up and trap the Sun's radiation in Earth's atmosphere. If the radiation cannot escape the atmosphere, temperatures rise.

Predictions

In an attempt to clarify the discussion, the Intergovernmental Panel on Climate Change, or IPCC (established in 1988 by the World Meteorological Organization and the United Nations Environment Programme), conducted an international assessment of global warming. The panel predicted a doubling of atmospheric carbon dioxide within the next century, and an associated average temperature rise of 1.5°–4.5°C. In response, governmental representatives from around the world met in Kyoto, Japan, to hammer out preemptive measures. The result was the Kyoto Protocol that, among other things, requires industrialized nations to reduce greenhouse gases at least 5 percent below 1990 levels by the period 2008–2012, and demands that funding be routed to the introduction of energy-saving technologies into developing countries (see Chapter 8, Documents, Letters, and Reports).

Predicted Consequences

If temperatures rise as predicted by the IPCC and numerous other reports, many biologists fear the effects will be widespread and serious. During the 1998 annual meeting of the American Association for the Advancement of Science, several well-known and respected scientists identified changes likely to result from such a global warming. Robert Watson of the World Bank was one of that number. He noted that the increase would represent the swiftest temperature rise on Earth in the last 10,000 years. Global warming at predicted rates will result in greater drought in the summer and in more precipitation in the winter, with the precipitation events becoming more severe—on the order of at least two inches of rain or two feet of snow at a time. Glaciers will shrink (more than half of the glacial mass in Glacier National Park has already disappeared in the past century). Sea levels will rise from 15–95 cm in the next hundred years, causing a loss of up to 30 percent of coastal wetlands. In response to these changes, Earth's species will shift. For example, the climate in the boreal forest will change to such a degree that two-thirds of its plant life will find its current location unsuitable. Forests normally advance or retreat by about 2–4 centimeters in a century—creeping into new areas as environmental conditions become favorable, or withdrawing from areas where conditions have become unhospitable. In a double-CO_2 world, these forests would have to move 150–600 km in a century to reach acceptable habitat, according to Watson. On the other hand, vector-borne diseases like malaria, dengue fever, and sleeping sickness will expand their territories, and invade new, once-colder areas.

Evidence

Scientists are now beginning to conduct experiments to determine the potential results of various levels of global warming.

One long-term study of the native blue grama grass, a major food of grazing animals in the prairies of Colorado, provided evidence that an increase of a few degrees in average low temperatures could lengthen the growing season enough to allow invasive grasses to take over. The study showed that a 1°C rise in average low temperatures equated to a one-third decrease in blue grama grass. Published in early 1999 by Richard D. Alward of Colorado State University, and others, it correlated the rise in global minimum temperatures, which is twice the increase in global maximums, with the shift in vegetation in the shortgrass steppe ecosystem they studied in Colorado. The steppe's dominant warm-season grass, *Bouteloua gracilis*, declined, while the cool-weather sedge, *Carex eleocharis*, and exotic forbs thrived. Based on their findings, the researchers cautioned that reduction in the warm-season grass could spark increased

A slight change in water temperature allowed mussels to dominate this ecosystem. Here, sea stars feed on mussels at the edge of a tidal pool.

Photo by Eric Sanford

invasion by exotic species and create an ecosystem that is less tolerant of drought and grazing.

Biologists are also noting ocean ecosystem changes associated with temperature variation. Field and laboratory experiments, reported in 1999 by zoologist Eric Sanford of Oregon State University, demonstrated that even small climatic changes could cause havoc in the relationships between predators and prey. Sanford's study focused on the sea star *Pisaster ochraceus*, a keystone predator, and the rocky intertidal mussels, *Mytilus californianus* and *M. trossulus*. The mussels are the sea star's primary prey. Temperature changes of 3°C slashed sea star predation by 29 percent. Based on these findings, the zoologist concluded that climatic shifts could have immediate and drastic consequences for ocean ecosystems.

Climate changes, and other weather phenomena like El Niño, may also be responsible for outbreaks of various diseases, such as malaria, cholera, and hantavirus pulmonary syndrome, according to a 1998 report from the American Academy of Microbiology. El Niño occurs when the tropical Pacific trade winds diminish, allowing a swath of warm water from Indonesia to flow eastward. The resulting warmer water and increased humidity in the air above it bring changes to weather patterns in North and South America and around the world. The report recommended establishment of a research project, the El Niño–Southern Oscillation (ENSO), and ensuing long-term studies to clarify the link between climate and infectious disease. It also identified a need for database networks so scientists could share their research findings more efficiently.

References

Alward, R., et al., "Grassland Vegetation Changes and Nocturnal Global Warming," *Science*, 8 January 1999: 229–31.

Colwell, R., and J. Patz, "Climate, Infectious Disease and Health: An Interdisciplinary Perspective," 1998, <http://www.asmusa.org/acasrc/aca1.htm>, accessed 2 February 2000.

Sanford, E., "Regulation of Keystone Predation by Small Changes in Ocean Temperature," *Science,* 26 March 1999: 2095–97.

Watson, R., "Climate Change," American Association for the Advancement of Science 1998 annual meeting, Philadelphia, 13 February 1998.

Carbon Sink

One phenomenon that has left scientists scratching their heads in the last decade is the whereabouts of the missing carbon sink. They have been unable to determine what happens to one-fourth—about 10^{12} kilograms—of the carbon released into the atmosphere each year as a result of human activities. The quandary gained heightened attention when the carbon cycle and the role of forests in that cycle became a key part of the Kyoto Protocol (see Chapter 8, Documents, Letters, and Reports). The treaty requires each country to assess its 1990 greenhouse gas emissions. The assessment will serve as a baseline to determine whether the country is living up to the agreement, which demands emission reductions of at least 5 percent below 1990 levels by the period 2008–2012. Knowledge of the role of forests is important, because scientists believe forests naturally take up huge quantities of the carbon dioxide humans discharge every year from automobiles and manufacturing centers. One 1999 report, however, had already removed the northern temperate forests from the list of candidates for the missing carbon sink.

Without a full understanding of the carbon cycle, the countries that signed the treaty will find it difficult to track their compliance and that of other nations. Scientists, including biologists, are now preparing and conducting a slew of experiments to measure carbon dioxide in and between the air and ecosystems on land and in water.

References

Kaiser, J., and Schmidt, K., "Closing the Carbon Circle," *Science,* 24 July 1998: 504.

Schindler, D., "The Mysterious Missing Sink," *Nature,* 11 March 1999: 105–07.

Schmidt, K., "Coming to Grips With the World's Greenhouse Gases," *Science,* 24 July 1998: 504–06.

CHAPTER FOUR
Evolution

The study of evolution has undergone a major expansion in recent years. Scientists now look to evolution not only to develop and refine the tree of life, but also to understand human behavior, to design new antibiotics, and even to help to solve crimes.

Biologists and other researchers are continuing to arrange taxa (species and groups of species), both extinct and extant, based on their similarities and differences. Many of these scientists, known as systematists, are now using cladistics to decipher evolutionary history. Cladistics involves a close examination of what these systematists call the *characters* present in an individual species or a group of species. A character can be something as obvious as the origin of feathers in an individual species, or something more subtle, like a change in vein pattern on a dragonfly wing or an alteration of a snake's venom. The systematists then make assumptions about the relatedness of the characters and group species accordingly.

Besides the shift toward cladistics, systematists have great challenges ahead. One exciting development is the entrance of genetics into evolutionary history. Geneticists are beginning to analyze the DNA of species to determine their relatedness. Their results often coincide with the conventional wisdom about species relationships, but not always. Some widely accepted relationships are now in doubt, and even the arrangement of the three major branches of life—the eukaryotes, the prokaryotes, and the archaea—are disputed.

Systematists are trying to weigh the relevance of genetics, morphology, behavior, and other factors in determining a species' location on the tree of life. Such information is becoming increasingly important as other scientists use these relationships in their work. For instance, details about the relationship of humans to other species is paramount in selecting an appropriate test animal (known as an "animal model") for the development of drugs.

Evolution also has a role in the mounting concern over antibiotic-resistant pathogens. As society has increasingly used antibiotics, pathogens have evolved defense mechanisms that allowed them to survive. Scientists are now scrambling to create new antibiotics that will circumvent those defenses.

Evolution is even joining the fray in the courtroom. A Louisiana judge recently allowed the use of phylogenies (evolutionary histories) in a case to determine the source of HIV transmission.

This chapter will describe some of the major findings in evolution over the past few years, including new ideas about the history and future of life on Earth.

References

Bull, H., and H. Wichman, "A Revolution in Evolution," *Science,* 25 September 1998: 1959.

Zug, G. R., *Herpetology: An Introductory Biology of Amphibians and Reptiles,* San Diego: Academic Press Inc., 1993: 326–27.

CONCEPTS IN EVOLUTION

Despite everything we've learned about evolution, it is still a field filled with unknowns: How did life begin? How does a species evolve? Can species evolve fast enough to outpace environmental change? Why are some groups of organisms so diverse, while others have only a few representatives? This section includes a brief look at some of the new concepts in the field and at some evidence for older views.

How Life Began

Scientists have long been debating the origin of life. (The "creationism" point of view is described in Chapter 6, Social Issues in Biology.) New studies are taking the birthplace of life from the warm, primordial pond of earlier hypotheses to the deep-sea vents, and these hypotheses are giving ammunition to those who believe life may not be restricted to Earth.

Several recent laboratory experiments have reportedly re-created the conditions found in hydrothermal vents on early Earth and have generated the precursors of life. In one experiment, led by Jay A. Brandes of the Carnegie Institution of Washington, D.C., scientists combined a mixture of water, nitrogen oxides, and sulfide minerals with the high temperatures and pressures of hydrothermal vents. Hydrothermal vents still exist on Earth today, and they are the source of some of the organisms now classified as Archaea, the recently added third branch of life. Within 15 minutes, the experiment produced ammonia. Next, the researchers added pyruvic acid to the experiment and soon procured the amino acid alanine, one of the building blocks of proteins. For the final step, they were able to link the amino acids into protein-like peptides.

Other research, reported in 1999 and led by Ei-ichi Imai of the Nagaoka University of Technology in Japan, showed that the rapid cooling and heating that occurs as materials flow into and out of hydrothermal vents is also conducive to the linking of amino acids.

These studies do not definitively preclude the primordial-soup hypothesis, however. More than four decades earlier, experiments that mimicked the much-less-severe conditions in a primitive pond also created alanine. On the other hand, those who favor hydrothermal vents as life's origination site believe the higher temperatures and pressures allow a greater number of reactions to occur faster, and that they therefore provide more opportunities for the formation of these precursors of life.

The work on hydrothermal vents was particularly heartening to those biologists and others who conjecture that life in this solar system may not be limited to Earth. They surmise that vents similar to those on Earth are likely present in the ice-covered oceans of Europa (one of Jupiter's moons) and possibly on other celestial bodies, and if so, they may present similar opportunities for life to arise.

Another study published in 1998 and led by Jean Chmielewski of Purdue University proclaimed that proteins—and not the nucleic acids DNA or RNA, as previously thought—may have sparked the origin of life. Chmielewski's research team developed a four-peptide system that was self-replicating and cross-replicating; both of these traits are believed to be prerequisites for life. In cross-replication, proteins and peptide fragments interact to make new molecules. The researchers believe their study is the first to show cross-replication in a peptide system, and it now throws peptides into the ring as candidates for being one of the initial keys to life.

References

Imai, E., et al., "Elongation of Oligopeptides in a Simulated Submarine Hydrothermal System, *Science*, 5 February 1999: 831–33.

Marshall Space Flight Center, "Clues to Possible Life on Europa May Lie Buried in Antarctic Ice," <http://www.jpl.nasa.gov/galileo/news11.html>, accessed 2 February 2000.

"Purdue Study Breathes New Life into Question of How Life Began," press release, 3 December 1998, <http://www.eurekalert.org/releases/pur-sbnliq.html>, accessed 2 February 2000.

Simpson, S., "Life's First Scalding Steps," *Science News*, 9 January 1999: 24–26.

Evolutionary Spurts

The fossil record, although sketchy at times, appears to show that some groups of animals appeared or diversified very rapidly. One of the most productive times was the Cambrian Period, from about 500 to 570 million years ago, which yielded most life as we know it today. Biologists and other scientists have long pondered the reasons behind this explosion of evolutionary activity. Hypotheses abound.

One hypothesis came from Australian biogeochemist Graham Logan and team members from both the Australian Geological Survey Organization and Indiana University. In their 1995 study, they suggested that the Earth's initial multicellular animals set up the animal explosion. These animals ate live plankton, and they competed against the bacteria, which had a diet of dead plankton. The bacteria dwindled, as did their oxygen-consuming lifestyle. With more oxygen available, other life forms were able to strengthen their meager toehold on life, and the species explosion began. Logan and his team based their hypothesis on carbon isotope data, which indicated that the bacteria declined at about the time of the Cambrian explosion.

Biologists also question how various taxa were (and are) prepared to evolve when the right conditions presented themselves. At a 1998 meeting, Molecular Strategies in Biological Evolution, scientists gathered in New York City to wade through various evidence and ideas and consider the possibility that random genetic mutations may not be the primary avenue of evolutionary change. The conventional wisdom held that a species evolves when a small, random, and rare mutation gives some individuals an advantage that has the ultimate effect of allowing them to reproduce more successfully. Generation after generation, the proportion of individuals carrying the mutation becomes greater. Eventually, the species incorporates the mutation; it evolves.

Now, however, many scientists believe the adjectives "small, random, and rare" should be replaced with "large, directed, and relatively common" in describing the mutations involved in evolution. For example, some genetic material may duplicate. While the initial material goes about its normal function, the duplicated genes can mutate, perhaps to take on new functions. In addition, some pieces of genetic material, like the so-called transposable elements, can jump from one location in the DNA to another and perhaps also take on new functions. Not only are duplications and transposable elements much more common than previously thought, some scientists believe that the cells can direct them. Instead of equating mutations with an unhealthy state, scientists now suspect that the genomes containing the mutation options are the ones most able to evolve fast enough to survive rapid changes in predation, food availability, or other environmental conditions.

According to one of the studies presented at the 1998 meeting, some parts of the genome may be able to mutate more quickly than others. Research on *Conus*, a genus of snails that use a multitude of toxins to secure their prey, showed that mutations in specific areas of genetic material generated the different toxins. According to the study, led by molecular biologist Baldomero Olivera of the University of Utah in Salt Lake City, the genomic flexibility appears to provide an advantage.

In another study published in 1998, researchers showed duplicated sequences of DNA in primates, including humans, and speculated that the duplications contained signals that instructed future copies of the duplications to travel to other parts of the genome.

Biologists are learning more with every study, and their work in these areas is far from over. Conjecture will continue on the important questions of how life began and how living things evolve.

References
Oliwenstein, L., "Life's Grand Explosions," *Discover,* January 1996: 42–43.
Pennisi, E., "How the Genome Readies Itself for Evolution," *Science,* 21 August 1998: 1131–34.

Co-evolution

The old notion of co-evolution has new evidence to back it up. First suggested several decades ago, co-evolution is the hypothesis that interacting species play off of each other's adaptations in such a way that they evolve in tandem. The idea has long seemed plausible, but supporting data have been difficult to obtain until recently.

A 1998 study of Midwest parsnip plants and webworms showed that these two species engaged in something akin to an unending version of the childhood card game called War. When the parsnips adjusted their store of defensive toxins, the webworms responded by modifying their attack tactics, so the parsnips again readjusted their toxins, the webworms again shifted their response, and so on. Entomologists May R. Berenbaum and Arthur R. Zangeri of the University of Illinois combed through a historical collection of parsnip plants dating to pre-webworm days and traced the amount of sphondin, a chemical compound that the webworms cannot metabolize. Prior to the webworms' arrival, the plants produced little or no sphondin. Current plant samples, however, contain high sphondin concentrations. In addition, the scientists found varying levels of sphondin and three other defensive compounds in the current plants, and similarly varying toxin resistance in the insects. Besides the evidence supporting the co-evolution concept, the work also demonstrated that evolution can be a rapid process.

Beetles and Plants

Another scientist, evolutionary entomologist Brian D. Farrell of Harvard University, went further back in time with his work to link the tremendous proliferation of beetles to the appearance of flowering plants, or angiosperms. The number of known species of Coleoptera, the beetles, is staggering. Estimates range from 290,000 to 330,000. Beetles represent nearly one-third of all known insects, and their diversity exceeds even that of the angiosperms by at least 17 percent. The unrivaled variety of beetles has given rise to a good deal of speculation about the reason behind it: Why is this one group of organisms so extremely diverse?

In this study, the researcher studied the herbivorous (plant-eating) beetles known as the Phytophaga, a group that makes up more than half of all beetles. After reconstructing the phylogeny of the beetles, based on 115 DNA sequences and 212 morphological characters, Farrell found that new lineages of angiosperm-feeding beetles correlated with times of enhanced rates of diversification. The emergence of the angiosperms provided an untapped resource for the taxa that were able to exploit it. The beetles began to feed on the plants, according to the study, and the phytophaga and the angiosperms co-evolved.

References

"Evidence Appears Strong to Bolster Concept of Co-Evolution, Scientists Say," University of Illinois at Urbana-Champaign, December 1998 <http://www.admin.uiuc.edu/NB/98.12/biotip.html>, accessed 2 February 2000.

Farrell, B., "'Inordinate Fondness' Explained: Why Are There So Many Beetles?" *Science*, 24 July 1998: 555–59.

Morell, V., "Earth's Unbounded Beetlemania Explained," *Science*, 24 July 1998: 501–02.

Wilson, E. O., *The Diversity of Life*, Cambridge, MA: Harvard University Press, 1993.

EXTINCTIONS

The history of life on Earth has been punctuated with a number of major extinctions. Some threatened to extinguish life altogether. So-called mass extinctions mark the delineations between many geologic time periods:

438 million years ago (MYA) at the end of the Silurian, 367 MYA to close out the Devonian, 248 MYA at the Permian's conclusion, 208 MYA at the end of the Triassic, and the best-known mass extinction of 65 MYA that marked the close of the Cretaceous and the demise of the dinosaurs. (A debate is currently under way over the placement of birds in evolutionary history as advanced dinosaurs.) Some scientists also believe we are currently in the midst of another mass extinction, this one brought about by humans.

Causes of Extinctions

Asteroids have become inexorably tied to extinctions in the public mind because of the popularity of dinosaurs and the well-publicized hypothesis that the collision of a giant asteroid with Earth triggered events that led to their extinction. Other, less-heralded reports blame asteroids for smaller extinctions, as well. A recent report, for example, points the finger at an asteroid or comet that smashed into Argentina 3.3 million years ago, leading to climate changes that caused the extinction of 35 mammals and a flightless bird.

While the causes of the mass extinctions always draw some debate, a number of recent studies have brought attention to smaller, more contemporary turning points in evolutionary time.

Effect of Humans

Back just a blink of an eye in geologic time—11,000 to 50,000 years—the last of an aggregation of great, wild beasts roamed the planet. Among them were 6-meter-long giant sloths, 7-meter-long monitor lizards, and 4.5-meter-tall mammoths. Then, they abruptly disappeared. Did climate change kill them off, or did humans?

A study published in 1999, led by Gifford H. Miller of the Institute of Arctic and Alpine Research and the University of Colorado, placed the blame for one and possibly more recent extinctions on humans. The research highlighted the extinction of an ostrich-sized, flightless bird known as *Genyornis newtoni*, which disappeared from Australia in the Late Pleistocene, along with most of the continent's other large terrestrial animals. The researchers surmounted one major hurdle to their work by pinpointing the time of extinction through a unique combination of techniques, including amino acid racemization, accelerator mass spectroscopy, and eggshell analyses using carbon 14 and thermal ionization mass spectrometry. Once the animal's demise was dated to 50,000 years ago, researchers could place humans, who arrived in Australia from 53,000 to 60,000 years ago, at the scene.

However, at the time of the bird's extinction, humans had not yet made their way to nearby New Zealand. New Zealand supported a thriving population of similar flightless birds. The close timing of human arrival in Australia and the *Genyornis's* departure, coupled with the presence of similar birds on a nearby island, led the researchers to implicate human activity—hunting, or possibly burning the landscape on which the bird relied for food—as the root cause of the extinction.

Climate Change

Climate and vegetation change were the culprits in the extirpation of North America's large grazing species, according to a presentation by paleontologist Steven Stanley of Johns Hopkins University at the 1998 meeting of the Ecological Society of America. About 6 million years ago, the continent lost most of its grazers, including rhinos and camels. He alleged that the climate-induced shift from wetter to drier weather, and from soft to tough grasses, brought about the change. Only animals with strong molars could grind up the tougher new grasses; the other species perished. Support for this hypothesis also came from evidence that, as the increasingly dry habitat lost its earlier inhabitants, a slice of wetter habitat persevered and continued to support softer grasses and its grazers.

Scientists may be able to learn more about extinct species' diets, and about the causes for extinctions, through an unlikely source: preserved dung. Molecular biologist Hendrik Poinar and geneticist Svante Pääbo of the University of Munich demonstrated how to procure usable DNA from coprolites, the technical term for preserved feces. The researchers discovered that the coprolites contained Maillard products, which are composed of protein, nucleic acids, and sugar. These Maillard products essentially coated and protected the DNA within. The researchers used a chemical called N-phenacylthiazolium bromide (PTB) to remove the coat and retrieve the DNA. Since most scientists now agree that amber is a poor preservative of DNA, despite the initial ballyhoo about its potential, scientists are now looking to coprolites as a promising store of genetic information. According to the 1998 report, fossilized droppings from an extinct ground sloth contained genetic material from both ingested food and from cells shed from the sloth's digestive tract. The DNA of the cells proved to closely match that of another species of ground sloth, which is also extinct. The DNA of the ingested food was traced to eight families of plants. The data gleaned from the sloth coprolites may also provide additional clues to the sloth's eventual disappearance.

Other scientists are expanding the analysis of fossilized excrement to Neanderthals. A discovery of coprolites in a Neanderthal-inhabited cave is giving researchers hope that they will soon obtain much-desired information about the place of these early humans in evolutionary history, as well as evidence about their diet.

References

Brown University, "Study Associates Asteroid or Comet Impact with Extinctions in Argentina," press release, 10 December 1998.

Flannery, T., "Debating Extinction," *Science*, 8 January 1999: 182–83.

Gibbons, A., "Ancient History," *Discover*, January 1998: 47.

Kaiser, J., "Tracking Vanishing Mammals and Elusive Nitrogen," *Science*, 28 August 1998: 1274.

Miller, G., et al., "Pleistocene Extinction of *Genyornis newtoni*: Human Impact on Australian Megafauna," *Science*, 8 January 1999: 205–08.

Stokstad, E., "A Fruitful Scoop for Ancient DNA," *Science*, 17 July 1998: 319–20.

Current Mass Extinction?

The magnitude of the mass extinctions listed at the beginning of this section were indeed great. Of those listed, the percentages of species that disappeared are

- 70 percent for the Cretaceous
- 79 percent for the Devonian and Triassic
- 84 percent for the Silurian (Ordovician)
- 95 percent for the Permian

At the 1999 meeting of the American Association for the Advancement of Science, ethnobotanist Paul Alan Cox of the National Tropical Botanical Garden said another mass extinction should be added to the list: that of the current era.

According to Cox, the mass extinctions share a number of traits. They represent a substantial loss of biodiversity over a taxonomically broad range, they are global, and they are rapid, meaning they occurred over a period of less than 10,000 years. The current situation on Earth meets all of these criteria, he said. Species are disappearing at a fast clip. The numbers of species that are threatened with extinction include

- 1,145 species (11 percent) of birds
- 5,064 species (11 percent) of mammals
- 190 species (3 percent) of reptiles
- 481 species (2 percent) of fish
- 244 species (30 percent) of gymnosperms
- 17,594 species (9 percent) of dicots
- 4,883 species (9 percent) of monocots

He also cited reports that one in eight plant species are threatened with extinction now, and if the pattern continues, half will be extinct within

30 years. He believes that this current period of extinction will be indistinguishable from past extinctions caused by an astronomical event, such as the hypothesized asteroid collision that dealt the deadly blow to dinosaurs.

Reference

Cox, P., "Conservation in Jurassic Park: Endangered Plants and the National Tropical Botanical Garden," topical lecture, American Association for the Advancement of Science 1999 annual meeting, Anaheim, CA, 22 January 1999.

NEW FINDINGS

New findings in evolution are coming quickly. This section provides a glimpse of the variety of the discoveries over the past few years.

Flatworm as Ancestor

For years, the Cambrian explosion—mentioned earlier in this chapter—has been a source of wonder as evolutionary biologists and others seek information about this period during which animals diversified at a staggering rate. A fundamental question is: Where did it all start?

A study reported in 1999 presented the modern-day animal that is most closely related to the organisms that started it all some 530 million years ago. According to the research team, which included geneticist Jaume Baguña and Iñaki Ruiz-Trillo of the University of Barcelona in Spain, the acoel flatworm is that modern-day animal. They believe this small, marine flatworm is a relic of Earth's first bilaterally symmetrical organisms and deserves its own spot in evolutionary history. Previous groupings placed it within the mainly parasitic Platyhelminthes.

Before bilateral symmetry, which gave animals distinct right and left halves, radial symmetry ruled. The shift to bilateral symmetry is seen as a key step in the evolution of higher organisms. After mounting a molecular investigation of 18 species of acoels, Baguña, molecular biologist Timothy Littlewood of the Natural History Museum in London, and others on the research team, compared the flatworms' genetic data with those of other species grouped in Platyhelminthes, and with those of other select organisms. The analysis showed that the acoels were strongly dissimilar to the other Platyhelminthes, and that they branched off the evolutionary tree somewhere between radial jellyfish and the first bilateral animals. The placement might also explain some of the other oddities of flatworms, such as their unusual cell-division pattern and lack of a larval stage.

In summary, from this research, the scientists inferred that the flatworms' ancestor could have been the spark that set off the Cambrian explosion.

References

Pennisi, E., "From a Flatworm, New Clues on Animal Origins," *Science,* 19 March 1999: 1823.

Ruiz-Trillo, I., "Acoel Flatworms: Earliest Extant Bilaterian Metazoans, Not Members of Platyhelminthes," *Science,* 19 March 1999: 1919–23.

Dinosaurs

Dinosaur Skin

Impressive ultraviolet-enhanced dinosaur fossils are now providing a look at the beasts' soft tissues and internal organs and are offering hints about their physiology. A team led by respiratory physiologist John A. Ruben of Oregon State University used ultraviolet imaging on fossils from a small raptor known as *Scipionyx samniticus.* The *Scipionyx* fossils included a nearly complete skeleton along with small remnants of soft tissues, including the liver, colon, and trachea. When the researchers illuminated these preserved tissues with ultraviolet light, more of the tissues became visible. The enhanced view led to recognition that the little meat-eaters probably were cold-blooded, and that their livers may have served as piston-like devices to allow their simple lungs to support periods of high metabolism.

A nearly perfect view of dinosaur skin made headlines in 1998 as a scientific team, including Brian G. Anderson of the Mesa Southwest

Impressions from the skin of a hadrosaur show dime-sized, ridged bumps.

Photo by Brian G. Anderson

Large ossified tendons from a hadrosaur, shown overlying skin impressions.

Photo by Brian G. Anderson

Museum in Arizona, unveiled impressions from the skin of a duck-billed hadrosaur found in New Mexico. The impression revealed skin with a series of dime-sized, ridged bumps.

At about the same time, another research team reported that they had discovered the first skin impressions from dinosaur embryos. Luis M. Chiappe of the American Museum of Natural History in New York, Rodolfo A. Coria of the Museo Municipal Carmen Funes in Argentina, and Lowell Dingus of the American Museum of Natural History and InfoQuest Foundation, were co-leaders of the expedition. The skin impressions showed a scaly surface similar to that of a modern lizard. The impressions came from a nesting site for sauropods, which were quadrupedal, herbivorous dinosaurs that grew to the tremendous length of 15 meters. Located in Argentina, the 70- to 90-million-year-old site contained thousands of eggs and dozens of embryos.

Quick Growth

While some scientists studied the skin of sauropod embryos, others reviewed their bones. One study, announced at the 1998 meeting of the Society of Vertebrate Paleontology and based on data collected by graduate student Kristina Curry of the State University of New York, indicated that the super-sized dinosaurs grew to their adult size in just 8–11 years, laying down an estimated 10.1 micrometers of new bone tissue every day for the duration of their growth. If they grew at the pace of modern-day reptiles, the sauropods would have required more than a century to gain full size. The sauropods' quick rate of growth isn't unique in the animal world—ducks, for instance, have a similar growth rate, but they quit growing after about three weeks. According to Curry's study, the sauropods apparently maintained the rate for a decade or more.

Fish-Eaters

In Niger, a group of biologists and other scientists has discovered a new genus and species of spinosaurids, a group of bipedal, fish-eating dinosaurs, and the fossils yielded clues about the group's evolution. The fossils, along with phylogenetic analysis, described an unusual dinosaur with an "enlarged thumb claw and robust forelimb (that evolved) before the elongated snout and other fish-eating adaptations in the skull." Led by Paul C. Sereno of the University of Chicago, the study also demonstrated that the new African species was related to a European member of the spinosaur family, indicating that some members of that family must have crossed the Tethys seaway, which separated what was then the land mass known as Pangaea.

Skeletal reconstruction of the spinosaurid *Suchomimus tenerensis*, based on preserved bones. (See also page 66.)

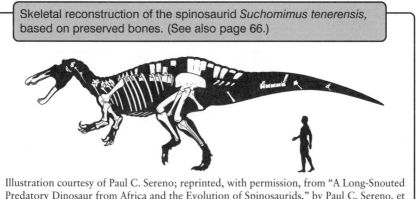

Illustration courtesy of Paul C. Sereno; reprinted, with permission, from "A Long-Snouted Predatory Dinosaur from Africa and the Evolution of Spinosaurids," by Paul C. Sereno, et al., *Science*, 13 November 1998: 1298–1302; ©1998 American Association for the Advancement of Science

References

American Museum of National History, "Unhatched Embryos Are First Ever Found of Giant Plant-Eating Dinosaurs," press release, 17 November 1998.

Monastersky, R., "Getting under a Dinosaur's Skin," *Science News*, 16 January 1999: 38.

Ruben, J., "Pulmonary Function and Metabolic Physiology of Theropod Dinosaurs," *Science*, 22 January 1999: 514–6.

Sereno, P., et al., "A Long-Snouted Predatory Dinosaur from Africa and the Evolution of Spinosaurids," *Science*, 13 November 1998: 1298–302.

Stokstad, E., "Young Dinos Grew Up Fast," *Science*, 23 October 1998: 603–4.

Wuethrich, B., "Stunning Fossil Shows Breath of a Dinosaur," *Science*, 22 January 1999: 468.

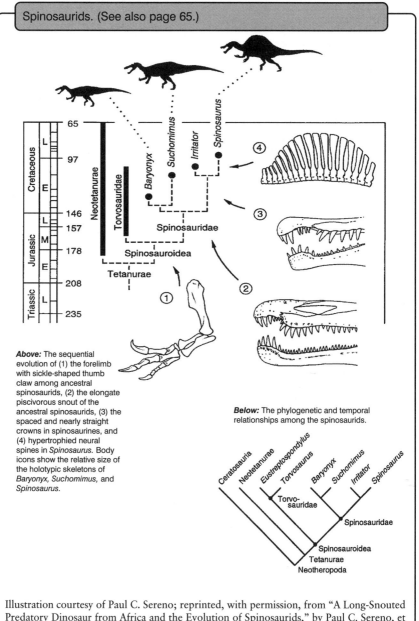

Spinosaurids. (See also page 65.)

Above: The sequential evolution of (1) the forelimb with sickle-shaped thumb claw among ancestral spinosaurids, (2) the elongate piscivorous snout of the ancestral spinosaurids, (3) the spaced and nearly straight crowns in spinosaurines, and (4) hypertrophied neural spines in *Spinosaurus*. Body icons show the relative size of the holotypic skeletons of *Baryonyx, Suchomimus,* and *Spinosaurus.*

Below: The phylogenetic and temporal relationships among the spinosaurids.

Illustration courtesy of Paul C. Sereno; reprinted, with permission, from "A Long-Snouted Predatory Dinosaur from Africa and the Evolution of Spinosaurids," by Paul C. Sereno, et al., *Science*, 13 November 1998: 1298–1302; ©1998 American Association for the Advancement of Science

Whales

Another group of large animals took the spotlight recently as researchers debated the origin and evolution of whales, or cetaceans.

Although scientists agree that whales evolved from mammals, they differ over which mammals are the whales' ancestors: the artiodactyls (which include pigs, camels, and cud-chewing animals like hippopotami) or the extinct, wolf-life mesonychians. Dental comparisons seemed to point to the mesonychians, but DNA evidence swung the pendulum to the artiodactyls.

In 1998, 50-million-year-old fossils of whale ankle bones added to the confusion. While some whale bones showed similarities with the artiodactyls, the ankle bone exhibited differences. At nearly the same time, paleontologist Hans Thewissen of Northeastern Ohio Universities College of Medicine reported that newly collected specimens of early whales and whale ancestors supported the whale-artiodactyl connection. The teeth of two of the whales were more primitive than those of the mesonychians, leading researchers to infer that whales could not be the mesonychians' descendants. Thewissen's data, however, were unable to definitively place the whales as a subgroup of the artiodactyls.

A marine biologist looked beyond bones and DNA to mother-taught behavior for clues to their evolution. Hal Whitehead of Dalhousie University in Halifax, Nova Scotia, suggested that the female adult sperm whale, and whales of other species, teach their young survival skills, which are passed from generation to generation. He based his work on differences between families, which he determined through genetic variation. Each family, he said, used distinctive whale songs. In other words, the female adults were passing down the songs from generation to generation. In addition, he suggested that survival skills, such as defense tactics, transfer to the next generation via the adult female. Whitehead acknowledged that the data aren't conclusive, but he felt they provided an intriguing possibility.

References

Normile, D., "Whale-Ungulate Link Strengthens," *Science,* 7 August 1998: 775.
Vogel, G., "DNA Suggests Cultural Traits Affect Whales' Evolution," *Science,* 27 November 1998: 1616.
Wong, K., "Cetacean Creation," *Scientific American,* January 1999: 26–30.

Turtles

Turtles, long pictured as the most primitive of reptiles, may require a new family photograph. A recent study by S. Blair Hedges and Laura L. Poling

Turtles may be a more advanced group than previously thought.

Photo by Ben Czinski

of the Institute of Molecular Evolutionary Genetics and Astrobiology Research Center in Pennsylvania added to the pool of evidence supporting an editing of the reptile phylogeny. The classical reptilian arrangement groups the crocodiles and birds in the Archosauria, and the tuataras, lizards, and snakes in the Lepidosauria. The turtles, on the other hand, sit in an ancestral position at the tree's base. The new work, however, turns that phylogeny nearly upside down, combining the turtles with the crocs, and relegating the squamates (lizards and snakes) to the primitive spot.

The identification of turtles as early reptiles arose because they were anapsids—they lacked the two temporal skull openings present in other reptiles, which are called diapsids. Researchers now claim that turtles are actually part of the diapsids, apparently evolving the anapsid condition secondarily. In addition, they hint that the tuataras may fit better with the archosaurs or turtles than with the squamates, but they admit that the evidence is still unclear.

The researchers based their conclusions on an analysis of nuclear genes and mitochondrial DNA. Mitochondrial DNA passes from mother to offspring in a genetically clear manner over time, thus providing reliable heredity information.

References

Hedges, S., and L. Poling, "A Molecular Phylogeny of Reptiles," *Science,* 12 February 1999: 998–1001.

Rieppel, O., "Turtle Origins," *Science,* 12 February 1999: 945–6.

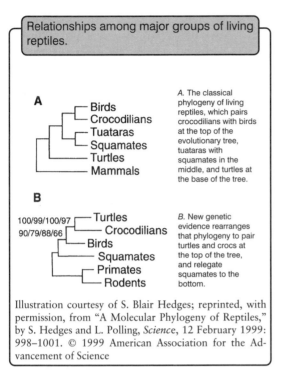

Relationships among major groups of living reptiles.

A
- Birds
- Crocodilians
- Tuataras
- Squamates
- Turtles
- Mammals

A. The classical phylogeny of living reptiles, which pairs crocodilians with birds at the top of the evolutionary tree, tuataras with squamates in the middle, and turtles at the base of the tree.

B

100/99/100/97
90/79/88/66
- Turtles
- Crocodilians
- Birds
- Squamates
- Primates
- Rodents

B. New genetic evidence rearranges that phylogeny to pair turtles and crocs at the top of the tree, and relegate squamates to the bottom.

Illustration courtesy of S. Blair Hedges; reprinted, with permission, from "A Molecular Phylogeny of Reptiles," by S. Hedges and L. Polling, *Science*, 12 February 1999: 998–1001. © 1999 American Association for the Advancement of Science

Humans
Footprints
Few paleontological findings can send tingles down the public's spine like human footprints dating back to the dawn of humankind. Paleontologist Lee Berger of the University of Witwatersrand in Johannesburg announced such a discovery in 1997. Geologist David Roberts of the Council for Geoscience in Cape Town found the footprints a year earlier near the Langebaan Lagoon in South Africa. The footprints, estimated to be 117,000 years old, were likely made by a woman or teenager who walked across a rain-soaked sand dune. From the distance between steps, researchers were able to estimate the individual's height as about five feet.

Within a year of the Langebaan Lagoon announcement, newly conducted dating pushed back the age estimate for a set of human footprints that had been found more than three decades earlier in Cape Town, south of the lagoon. Roberts, who discovered the Langebaan footprints, led the project. Using thermoluminescence, which reveals when sunlight last struck underlying sediment, the scientists pronounced the footprints to be 200,000 years old.

Out of Africa?
Like a teeter-totter on a child's playground, the debate on the geographical origin of humans has tipped back and forth, but with extra weight on what has been dubbed the out-of-Africa hypothesis. According to this view, humans arose from a small population of females in Africa. This population has gained the nickname "mitochondrial Eve." Another popular theory, the multiregional hypothesis, holds that humans from various regions of Africa, Asia, and Europe interbred and together evolved as one species. The differences of opinion arise over the analyses and conclusions drawn from genetic studies.

In 1997, for example, geneticist Michael Hammer of the University of Arizona analyzed the section of DNA called YAP that clearly traced the lines of heredity by way of the father-to-son transmission of the Y chromosome. Based on his findings, he reported that some modern African men have DNA descended from Asian men. In other words, populations of early humans migrated from Africa to Asia, where they remained long enough put their own identifying mark on the DNA of people in that region, and then returned to Africa.

Another study, published in 1999 and led by population geneticist Jody Hey and anthropologist Eugene Harris of Rutgers University, took a swing at both the out-of-Africa and the multiregional hypotheses. These researchers asserted that not one population, but two, gave rise to modern-day humans. The family tree for humans, then, would have two distinct trunks, one for Africans and one for non-Africans. They based their conclusions on an investigation of a single gene from different, current-day human populations. Using the variance in the gene from population to population, they constructed a history of the gene, and traced it to two ancestral gene versions, or haplotypes. Hey reported that they also found a site in the DNA—one base—that is different in Africans compared to non-Africans, the first time such a difference has been identified in humans, she noted.

So far, scientists within this field, which has been newly named anthropological genetics, only agree that they will be sorting through uncertainties and new studies for some time before the origin of humans is summarily settled.

References
Bower, B., "DNA's Evolutionary Dilemma," *Science News*, 6 February 1999: 88–90.
Holden, C., ed., "Humanity's Baby Steps," *Science*, 27 November 1998: 1635.
Menon, S., "Footprints from the Human Dawn," *Discover*, January 1998: 34.
Pennisi, E., "Genetic Study Shakes Up Out of Africa Theory," *Science*, 19 March 1999: 1828.
Svitil, K., "Out of Africa and Back," *Discover*, January 1998: 34.

CHAPTER FIVE
Molecular Biology and Genetics

I n the early twentieth century, the work of Gregor Mendel (1822–84) formed the foundation of genetics. Through his experiments with garden peas, Mendel described a mathematical pattern to demonstrate how individual differences—single traits—were passed from parent to offspring. His work remained largely unnoticed until three botanists rediscovered Mendel's paper in 1900 and helped to expand his ideas into the field of genetics. With the advent of molecular biology and genetics, biologists gained a whole new window on the living things they have studied for so long. By peering into the cells of organisms, dissecting the mechanisms and functions of the molecules inside, and unraveling the genetic code, biologists are gaining an understanding of the very basis of life. At the same time, they are marveling at the possibilities for the future.

Molecular biology and genetics touch on many other disciplines, as well. In evolution, for instance, knowledge of the genetic makeup of different organisms will help to refine evolutionary lineages and will likely cause at least a few reorganizations. Environmental scientists, too, may be able to tap genetic information to discover useful details about the life history of an invading species, and then use that information to curtail the species' spread. This chapter gives an overview of some of the important work done in molecular biology and genetics in the mid- and late 1990s.

READING THE GENETIC CODE

One of the most rapidly advancing areas of study is the reading of an organism's full genetic code, or genome. Not long ago, the idea that science would someday be able to sequence—that is, to determine base by base, or genetic letter by genetic letter—the entire genome of a living thing was no more than a dream. New technology, however, teamed up with ingenuity and unwavering ambition to turn the tide. In 1995, for the first time, scientists sequenced the genome of an organism. Since then, the number of sequenced organisms has blossomed, with scientists estimating that the human genome will be unraveled very close to the beginning of the twenty-first century.

A History

The First Sequence

J. Craig Venter and researchers at his Institute for Genomic Research in Maryland became the first to sequence an organism's genome. In July 1995, using whole-genome shotgun sequencing, a technique that had previously won little support, they deciphered the genetic code of the bacterium *Haemophilus influenzae*. The genome contained 1.8 million base pairs in its lone circular chromosome.

Before the sequencing of *H. influenzae*, scientists felt the only way to unravel the genetic code of an organism was to create an initial map that would guide the placement of the genetic information as it was sequenced. For *H. influenzae*, however, the Venter research team broke up millions of copies of the single chromosome into short strands of random lengths, then sequenced the strands. Because the strands had different starting and ending points, they were able to find overlapping sequences and built the genome by matching them up. Matching the pieces was difficult and time-consuming until institute researchers developed software to speed the process.

Three months after announcing the sequence for this bacterium, Venter's team deciphered the genome of another bacterium, *Mycoplasma genitalium*.

Bacteria, Pathogens and Others

By the beginning of 1999, less than four years after the initial genetic sequence of *H. influenzae*, scientists had deciphered the genomes of many other organisms. These included the bacteria *Escherichia coli* and *Bacillus subtilis*, the former found in the human digestive tract and the latter in soil. Both topped the 4-million-base-pair (4-megabase) mark. Other resolved genetic codes included the ulcer-causing *Helicobacter pylori*; the organisms responsible for Lyme disease, syphilis, and tuberculosis; a yeast; and several archaea from the newly added branch on the evolutionary tree.

The existence of described genomes for these organisms presents a multitude of possibilities for research. *Chlamydia trachomatis* is an example. Sequenced in 1998 by Richard S. Stephens of the University of California–Berkeley and a team of researchers from several institutions, the bacterium is a major cause of preventable blindness and is also responsible for a common genital tract infection. With a blueprint of its 1.0-million-base-pair genome, researchers can now begin to understand how the bacterium grows and lives within its host eukaryotic cells, which perhaps will enable them to develop new preventive approaches or treatment options. The genomic data also indicated that the bacterium has acquired five or more times the number of eukaryotic genes as other bacteria have, presenting another puzzle worthy of investigation.

The 1.6-million-base-pair sequence of *Campylobacter jejuni* also opens doors for researchers. Bart Barrell and Julian Parkhill at the Sanger Centre in Cambridge, U.K., sequenced the genome. The food poisoning caused by this bacterium can lead to the neuromuscular disorder named Guillain-Barré syndrome. The DNA sequence will provide added avenues of exploration in linking the bacterium with the resulting disorder.

First Multicellular Organism

While 1998 brought the sequences of the microorganisms behind syphilis, tuberculosis, and *Chlamydia*, the spotlight fell most brightly on a nematode. *Caenorhabditis elegans* became the first multicellular organism to have its genome sequenced.

Large research groups at Washington University in the United States and the Sanger Centre in the United Kingdom cooperated with each other and with other researchers to sequence the 97-million-base (97-megabase) genome of the nematode, a spindle-shaped roundworm. Even before its genetic makeup was revealed, *C. elegans* was a well-described animal. With less than 1,000 somatic (non-reproductive) cells, including only 300 neurons, researchers had already studied nearly every nook and cranny on and in the nematode. The creature is also transparent, which literally provided a window to its development. Its genetic mutations and their results were also well known.

To arrive at the sequence, the research groups built upon the physical map of the genome, which was devised by John Sulston and Alan Coulson of the Medical Research Council Laboratory of Molecular Biology in Cambridge. Sulston and Coulson turned to the fairly new technique of using yeast artificial chromosomes, or YACs, to fill in the gaps on their map. With the completed map, a research group could then separately sequence each portion of each chromosome, and insert that sequence into the proper place on the map. The challenge was great, considering that only three years earlier, the scientific world had heralded the very first sequence of an organism, a bacterium that was less than one-fifth the size of *C. elegans*. Undaunted, the research groups pushed forward and published a virtually complete sequence of the first animal genome.

The work was important on a number of levels. While the functions of some animal genes were known before the nematode was sequenced, the functions of many others were not. The complete genome yielded the information needed to trace each gene's role. An understanding of those roles could be extended to other multicellular organisms, including humans. In comparing proteins from *C. elegans* with other organisms, the researchers found that the nematode's proteins matched those of the other species in the following percentages: *E. coli*, 9.1 percent; the yeast *Saccharomyces cerevisiae*, 26 percent; and humans, 36 percent. This finding reflects the evolutionary relationships of those organisms.

The results of this work and other genome sequencing projects will have ramifications for years to come. The genome will become a reference point for future studies on the nematode, on organisms in general, and on animals in particular. It may also give scientists insights into the evolutionary shift from unicellular to multicellular organisms. In addition, the work leading to the *C. elegans* sequence blazed a path for other genome research projects, including the Human Genome Project.

References

C. elegans Sequencing Consortium, "Genome Sequence of the Nematode *C. elegans*: A Platform for Investigating Biology," 11 December 1998, *Science:* 2012–18.

"Genomes Galore," *Science,* 19 December 1997: 2042.

Hatch, T., "*Chlamydia*: Old Ideas Crushed, New Mysteries Bared," *Science,* 23 October 1998: 638–39.

Oliwenstein, L., "Bacterium Tells All, Human Tells a Lot," *Discover,* January 1996: 32–33.

Pennisi, E., "First Food-Borne Pathogen Sequenced," *Science,* 26 February 1999: 1243.

———, "Worming Secrets From the *C. elegans* Genome," *Science,* 11 December 1998: 1972–74.

Plasterk, R., "Hershey Heaven and *Caenorhabditis elegans*," *Nature Genetics*, January 1999: 63–64.

Stephens, R., et al., "Genome Sequence of an Obligate Intracellular Pathogen of Humans: *Chlamydia trachomatis*," *Science,* 23 October 1998: 754–59.

Human Genome Project

An endeavor of monumental proportions, the Human Genome Project has touched research labs around the world. The deadline for the international project, which has the goal of sequencing the approximately 3 billion bases in the human genome, has grown earlier and earlier as funding has increased, techniques have improved, and competition from private ventures has become more intense. A completion date of spring 2000 was expected.

The race to sequence the human genome heated up tremendously in the spring of 1998 when scientist J. Craig Venter announced that he and his new company, Celera Genomics, planned to sequence the human genome by 2001. Venter had already proven his genome-sequencing abilities in 1995 when he and his Institute for Genomic Research, a nonprofit affiliate of a for-profit company, became the first to decipher the genetic code of an organism, the bacterium *Haemophilus influenzae.* Venter's bold declaration piqued the interest of many researchers, who feared that Celera planned to patent the genetic code.

Although a patent would not give the company ownership of DNA in its natural state, the company could potentially patent the process it used to build its database. Under U.S. patent laws, a company could patent both the process and the database of information generated by that process. If another research group then wished to access the patented database, it would have to go through the company. The research group's other option would be to gather the information itself using a different, nonpatented method. The vast majority of researchers, however, have neither the time nor the funding to take that avenue.

With the threat of genome patents looming in the wings, the U.S. government's National Human Genome Research Institute and the Wellcome Trust of the United Kingdom began essentially to race against Venter and his company. They boosted the Human Genome Project with large grants for four groups: the Sanger Centre in Cambridge, the Robert Waterston group at Washington University, the Richard Gibbs group at Baylor College of Medicine in Texas, and the Eric Lander group at the Whitehead Institute for Biomedical Research in Massachusetts. They, along with the Joint Genome Institute, which is part of the U.S. Department of Energy, expressed confidence that they would have 90 percent or more of the genome sequenced by March 2000, followed by a two- to three-year period for a data review and ultimate presentation of the entire genome.

To speed the process, the research teams are employing automated DNA sequencers. The Sanger Centre reported in the 19 March 1999 issue of *Science* that its combined equipment could sequence a maximum of 32,000 samples—or fragments of DNA—each day. The equipment of choice includes DNA sequencers that can decipher the longest DNA fragments in the shortest time. The researchers prefer the longer fragments, because they can more easily find their correct location along the 3-billion-base length of the human genome.

References

Marshall, E., "NIH to Produce a 'Working Draft' of the Genome by 2001," *Science*, 18 September 1998: 1774–5.

Mullikin, J., and A. McMurray, "Sequencing the Genome, Fast," *Science*, 19 March 1999: 1867–8.

Pennisi, E., "Academic Sequencers Challenge Celera in a Spring to the Finish," *Science*, 19 March 1999: 1822–3.

Wertz, D., "Patenting DNA: A Primer," *Gene Letter*, February 1999, <http://www.genesage.com/professionals/geneletter/archives/dna1.html>, accessed 30 March 2000.

DNA CHIPS

One technological advance that is greatly expanding the applications for molecular biology studies is the DNA chip. Biology labs and medical clinics will both likely contain a collection of DNA chips in the future. Many already use them. With the outward appearance of a computer chip, the DNA chips have replaced the transistors with strips of DNA. The DNA chips, also known as arrays, are designed to quickly detect when a cell or virus exactly matches one of the tens of thousands of different DNA strips on the chip. In the medical clinic, the chips can serve as diagnostic tools. In the biology lab, they can provide researchers with nearly instant genetic information about mutations in a test organism.

How They Work

The DNA chips work by allowing the DNA under investigation to bind strongly to one of the DNA strips contained on the chip. DNA contains the four bases cytosine, guanine, adenine, and thymine, commonly referred to as C, G, A, and T. Cytosine on one strand links up with guanine on the other. The same match occurs between adenine and thymine. In other words, a strand of DNA containing an adenine-thymine-adenine (ATA) sequence would match with a complementary TAT sequence. Strongly bound DNA snippets are indicated by an intense signal at the location of the DNA strip, or probe. Because the researchers know the composition of the strip (say, ATAC), they can easily deduce the exact genetic sequence of the DNA that strongly binds to it (TATG). Scientists can, of course, use other methods of DNA sequencing, such as those mentioned in the previous section, but they are much slower.

An obvious benefit from this technology is the chip's ability to detect genetic disorders. If a chip is designed with DNA strips containing the mutations that are suspected of playing a role in various disorders, such as Alzheimer's disease or certain types of cancer, the chip could be used to quickly survey a patient's genome for those mutations. Another chip that held different viral mutations could determine the exact variant of a virus in an infected patient. With that information, physicians might be able to prescribe the best treatments for that variant.

Whole Genomes

One team of researchers used DNA chips to sequence the whole genome of an organism: yeast. Using a chip that contained DNA strips from previously sequenced yeast, the research team tested the DNA from two other yeast strains, one of which was almost identical to that used to make the chip. The two strains were also different in that one carried a drug resistance. While fragments from both strains bound to many of the strips, each strain had some fragments that bound to unique strips, according to Elizabeth Winzeler, Dan Richards, Ronald Davis, and their colleagues at Stanford University in California, Duke University in North Carolina, and the Affymetrix firm in California. The nearly identical yeast strains bound as expected, but the second, drug-resistant strain contained a number of snippets that did not exactly match the DNA strips, indicating that its genome contained base differences. By conducting further comparisons, the researchers were able to identify the region of the DNA that conferred the drug resistance. This work with yeast may prove to be a step toward perfectly tailoring drugs to fit an infection.

DNA chips are just entering labs and clinics. The clinics are using them mainly to detect HIV mutations. As the technology improves and the word gets out, they may well become a fixture in the labs and clinics of the future.

References

Service, R., "DNA Chips Survey an Entire Genome," *Science*, 21 August 1998: 1122.
Wickelgren, I., "DNA Chips: Here Come Genetic Diagnoses, at Pentium Speed, for a Host of Diseases," *Physician's Weekly*, 17 November 1997, <http://www.physweekly.com/archive/97/11_17_97/twf.html>, accessed 2 February 2000.

GENETIC ENGINEERING

Genetic engineering is a term applied to the process of altering the genetic complement. For thousands of years, humans have been orchestrating genetic alterations via selective breeding, which is the method of breeding plants or animals for specific traits deemed desirable by the person directing the reproductive project. A plant hybridizer, for instance, might breed a rose for color, shape, disease resistance, or a variety of other traits. Now, the term genetic engineering has been used to describe the process of transferring genetic material from one organism into another. The practice has also generated intense debate about whether the creation of recombinant organisms is wise (see Chapter 6, Social Issues in Biology).

Crops

Some scientists believe that the genetic engineering of plants is the only way to enhance crop productivity to a degree sufficient to meet the demands of the growing human population. While the average yields of some crops, including maize and possibly other cereals, have increased annually, their maximum yields have not. That dichotomy suggests that farmers are becoming better at growing their crops, but the crops' potentials are capped. With an estimated 40 percent rise in rice, wheat, and maize demand by the year 2020, many scientists and economists wonder whether current cereal plants will fulfill the need. Other scientists, however, question the crop ceilings and insist supply will meet demand.

While discussion continues, scientists are creating genetically engineered produce, many types of which are already available for purchase in grocery stores. Some efforts are directed toward increasing the density of plants a field can support, others hope to increase the size or number of vegetables per plant, and still others strive for disease- or pest-resistant varieties. Many recombinant vegetables have already received U.S. government approval, including potatoes, soybeans, corn, and tomatoes, and others are on the horizon.

Foods aren't the only crops selected for genetic engineering. In 1996, researchers Maliyakal E. John and Greg Keller at Agracetus, located in Wisconsin, added genes to cotton seedlings to develop plants that could make a high-performance fiber combining the traits of cotton and those of the polyester polyhydroxybutyrate (PHB). Although plastics are considered purely synthetic, most are strings of organic molecules. The method of delivery they used was a gene gun, which shot into the seedlings beads carrying the gene for PHB. While far from foolproof, John and Keller were able to create 30 mutant plants that produced some PHB within the cotton fibers. To call the experiment an economic success, however, they needed to generate plants that produced at least three times as much PHB as the initial mutants. Subsequent work on PHBs by other researchers was less successful than hoped, and some have begun studying a different version of PHB.

References

Mann, C., "Crop Scientists Seek a New Revolution," *Science,* 15 January 1999: 310–14.

Moffat, A., "Toting Up the Early Harvest of Transgenic Plants," *Science,* 18 December 1988: 2176–8.

Moore, S., "Natural Synthetics," *Scientific American*, February 1997: 36–37.

Pests

In the fight against pests, particularly against insects, scientists are also turning to genetic engineering. Mosquitoes have become a prime target.

Mosquitoes

Researchers, including Jian-Wei Liu of the National University of Singapore, set their sights on two genes of the bacterium *Bacillus sphaericus*. The bacterium is lethal to the two types of mosquitoes whose bites can cause malaria, St. Louis encephalitis, and elephantiasis. In the mid-1990s, Liu's research team began inserting the *B. sphaericus* genes into another bacterium called *Asticcacaulis excentricus*, which then became toxic to the mosquitoes as well. The advantage of the recombinant *A. excentricus* was that in laboratory conditions, it reproduced more easily, and less expensively, than the *B. sphaericus*. In addition, the recombinant bacterium did not have the enzymes that eventually degrade the toxins, unlike the *B. sphaericus* bacterium.

At about the same time that Liu's team was conducting its research, two scientists from the University of Memphis, S. Edward Stevens Jr. and Randy Murphy, were inserting the genes of *Bacillus thuringiensis israelensis* (Bti), a bacterium that is toxic to both mosquitoes and black flies, into blue-green algae. The alga became a vector for the Bti poisons.

The battle against mosquitoes saw another advance in 1998 when two labs worked together to show that they could genetically engineer mosquitoes to pass down an added gene from generation to generation. The labs of Anthony A. James of the University of California–Irvine and Frank H. Collins of Notre Dame University in Indiana conducted their research with *Aedes aegypti*, the mosquito that transmits both the seldom-fatal dengue fever, and the more dangerous and often life-threatening yellow fever. The researchers were successful in inserting a gene for eye color into the mosquitoes, then waited to see whether the offsprings' eyes would be the naturally occurring white or the new reddish color. Those born with red eyes have since passed down that trait to subsequent generations. The work is a step in the right direction for other researchers who want change the genetic makeup of mosquitoes so that the insects can no longer transmit disease-causing parasites and viruses. If a new trait is heritable, scientists may be able to release the genetically engineered mosquitoes to the wild as a first step toward spreading the trait throughout the species.

References

Adler, T., "Mauling Mosquitoes Naturally: New Ways to Silence the Buzz," *Science News*, 27 April 1996: 270.

Travis, J., "Colorful Gene Marks Mosquito Manipulation," *Science News*, 4 April 1998: 213.

Medicines?

Genetically engineered animals and plants are also now beginning to produce a variety of pharmaceuticals, including drugs used in the treatment of diabetes, cystic fibrosis, and coronary-bypass patients. Like the genetically engineered crops mentioned previously, scientists are inserting genes for desirable traits. Here, however, they are adding genes that will produce human antibodies, enzymes, and disease-fighting drugs.

Vaccines

Led by William H. R. Langridge, a research group at Loma Linda University School of Medicine in California is creating potatoes that can make the vaccine for cholera. The inserted gene generates a protein found in the cholera toxin. Following the standard vaccine process, the protein, while not toxic itself, serves as the catalyst for the human body to make antibodies. If a person contracts the illness later, the body already has the antibodies to fight it off. According to the research team, cooking the engineered potatoes diminishes the potency a bit, but a person still would have to eat only one cooked, genetically modified potato a week for a month to receive full immunization. Instead of booster shots, the person would occasionally eat a potato to maintain that level of protection.

Another research group in 1998 also reported using potatoes in their work, but instead of a cholera vaccine, they inserted the gene to confer immunity to *E. coli*. They are planning clinical trials.

Antibodies

Various companies are also engineering plants to carry human antibodies. One drug company has already begun testing corn-produced antibodies as a treatment vector for radioisotopes, which would attack and destroy tumor cells in cancer patients. Other products include soybean-produced antibodies to fight herpes simplex virus 2, and tobacco-generated antibodies to fight tooth decay. On another front, the Loma Linda group has reported a potential for edible insulin by way of transgenic plants.

Transgenic Animals

Genetically engineered animals are also becoming drug producers. One such sheep-produced drug, alpha-1-antitrypsin, shows promise as a treatment for people with cystic fibrosis. This drug, a protein, stops the degradation of connective tissue that is associated with the disease. Developed in part by the same group of researchers who cloned Dolly (see Cloning, below), the genetically engineered sheep yields the protein through her milk. She has passed along the ability to make alpha-1-antitrypsin to her offspring, some of whom are producing milk with up to 17 grams of the substance per liter.

Research in 1998 gave a boost to the field of genetic engineering in animals. A research team announced that the percentage of successful gene transfers increased sharply when they inserted genes earlier in a cow's development. By adding the gene to the eggs before fertilization, virtually all of the cells of the resulting animal carried the new gene. Previously, the standard method was to insert the gene later in the cow's development. Because not all of the cells received the new gene, only the progeny of the engineered cells carried the new gene. The new method increased the gene-transfer efficiency from 1–10 percent to virtually 100 percent.

References

Gibbs, W., "Plantibodies," *Scientific American*, November 1997: 44.

Moffat, A., "Down on the Animal Pharm," *Science,* 18 December 1988: 2177.

Raloff, J., "Taters for Tots Provide an Edible Vaccine," *Science News,* 7 March 1998: 149.

TISSUE ENGINEERING

Another type of "engineering" with biological ties is tissue engineering. Here, the engineered tissues are grown in a lab instead of on or in a living organism. Already, several engineered tissues, including skin and cartilage products, have made it from the drawing board to the marketplace—or, more specifically, to the medical clinic. The engineered tissues will serve as replacement tissues to help people heal from injuries or other damage, and may one day provide a source of organs for transplants.

Tissues

In 1997, the U.S. Food and Drug Administration (FDA) gave its seal of approval to two skin products, TransCyte and Apligraf, and a cartilage product, Carticel. TransCyte, developed by Advanced Tissue Sciences Inc., in California, is composed of a polymer and cells from the inner layer of the skin. Apligraf, developed by Organogenesis Inc., in Massachusetts, contains both the inner and outer layers of skin—the dermis and epidermis. The FDA approvals allow both skin products to be used to promote healing: TransCyte to be layered over burns as a temporary fix, and Apligraf for wounds caused by venous ulcers in the legs. Carticel, made by Genzyme Corp., of Massachusetts, is actually cartilage grown in the lab from the patient's own cartilage-generating cells. The resulting tissue is used to replace the patient's damaged knee cartilage.

In the case of Apligraf, the artificial skin's development was based on research showing that the dermis-generating fibroblast cells would grow on a scaffold made of a collagen gel, and that the epidermis-generating keratinocytes would grow on the resulting dermal sheet. Because other companies held the patents to the methods for culturing the keratinocytes, Organogenesis developed its own technique using fibroblasts and keratinocytes gathered from the newly circumcised foreskins of newborns. Clinical trials of the bilayered, artificial skin on venous ulcers were encouraging, according to Nancy Parenteau, Organogenesis's chief scientific officer and senior vice president for research and development. In an article she wrote for *Scientific American*, Parenteau said those ulcers that had been the most difficult to treat were also the ulcers that benefited most from Apligraf: After 24 weeks, 47 percent of Apligraf-covered ulcers were completed closed, compared to 19 percent of conventionally treated ulcers.

While these products make their way into medical use, other companies are at various stages of research in engineering tissues to serve as bladder and heart valves, to treat foot ulcers in people with diabetes, and even to replace intestinal, corneal, and breast tissue.

References
Ferber, D., "From the Lab to the Clinic," *Science*, 16 April 1999: 423.
Parenteau, N., "The Organogenesis Story," *Scientific American*, April 1999: 83–84.

Organs
While artificial skin and cartilage have already arrived in the medical setting, researchers are struggling to create laboratory-grown organs. In the meantime, the demand for transplant organs is increasing with the aging human population.

Researchers are now building on the successes of engineered skin and cartilage, taking the first steps toward living, artificial organs, known as neo-organs. One research group, led by Anthony Atala of the Children's Hospital, Boston, has grown and transplanted urinary bladders into dogs. For their work, Atala's group grew the two components of bladders—the muscle cells and the urothelial cells that line the bladder and contain the urine—in a special combination of growth factors and on the right polymer scaffolding. They used biodegradable plastic molds of the bladder to shape the growing tissue into the bladder, with the muscle on the outside and the urothelial cells on the inside. They then transplanted the new bladders into the dogs, and found that they began to work normally within a month. After the plastic mold degraded, the bladders continued to function normally. Although the leap from dog bladders to human bladders will involve a good deal of research, this research is providing important information for other labs striving to make engineered organs for humans.

Biomedical engineers Robert Langer of the Massachusetts Institute of Technology, and Laura Niklason of Duke University in North Carolina, along with Jinming Gao, now of Case Western Reserve University in Ohio, had similar success growing pig arteries. The potential exists to take advantage of similar techniques to grow human arteries for use during coronary bypass operations.

All of the work combined sheds light on the future of tissue engineering and organ transplantation.

References

Ferber, D., "Lab-Grown Organs Begin to Take Shape," *Science*, 16 April 1999: 422–5.

Mooney, D., and A. Mikos, "Growing New Organs," *Scientific American*, April 1999: 60–65.

Parenteau, N., "The Organogenesis Story," *Scientific American*, April 1999: 83–84.

CLONING

While cloning has been a part of science fiction books and movies for years, researchers only recently cloned the first adult mammal, a sheep. Also unlike the movies, the clone was not an immediate and exact replica of the adult ewe. Nonetheless, the announcement sent a shiver of excitement—and, in some cases, fear—down the spines of scientists and lay people alike. (A discussion about the concerns surrounding cloning is included in Chapter 6, Social Issues in Biology.)

Dolly

In 1997, a team of researchers from Scotland's Roslin Institute and PPL Therapeutics stunned the world with the announcement that they had cloned an adult sheep to produce a lamb they named Dolly. Cloning itself was nothing new to scientists. Animal breeders had been cloning animals for decades by splitting embryos to create genetically identical siblings. Similarly, scientists had long known how to replace the DNA-containing nucleus of a unfertilized egg with the nucleus of a donor cell from a early-stage embryo. Both of these methods produced clones: genetically identical organisms. What was different about Dolly was that she marked the first time the nucleus of an adult mammal cell had been used to produce a viable clone.

The Idea

The Dolly project began as Keith H. S. Campbell of PPL Therapeutics worked with the Roslin Institute's Ian Wilmut and a team of researchers to review the successes and failures of previous adult-cloning experiments. The research team hypothesized that a successful cloning might best begin with an adult cell that had ceased growing. An adult cell whose genes were inactive would more closely mimic the situation in an embry-

Dolly.

Courtesy of the Roslin Institute

onic cell and might present the right conditions for the adult-cell cloning. After all, previous work had already shown that an embryonic cell could be successfully cloned. Past research had also indicated that the chromosomal material (or chromatin) within the nucleus of an adult cell is different than that of an egg or embryo cell. The difference appeared to be responsible for the inability to clone an adult nucleus.

The Method

Wilmut's research group induced inactivity in the adult cell by starving it. They believed the resulting nucleus would engage in some changes within its chromatin and become more conducive to totipotency, the all-important ability to differentiate into different types of cells. Totipotency is seen in embryonic cells, which differentiate into all of the cell types ultimately present in an adult. The researchers were right, but their experiments were not immediately fruitful. Of 277 donor-cell attempts, only Dolly was successful and made headlines.

Following the announcement of the project's success, many scientists noted that Dolly wasn't genetically identical to the nucleus-donating sheep, because she received the DNA only from the donor cell—a mammary cell from a 6-year-old Finn Dorset ewe—and not the bit of genetic material contained in her mother's mitochondrial cell. Many scientists, including Wilmut, also questioned whether the use of six-year-old DNA would have any consequences for the cloned sheep. They will be watching her closely as she ages to spot any effects of the "old" DNA. A Poll Dorset sheep provided the enucleated cell, along with the mitochondrial DNA. A sheep from the Scottish Blackface breed served as the surrogate mother. The use of three breeds helped confirm that Dolly was, in fact, the clone of the nucleus donor. Despite her not-quite-identical status, the debut of the seven-month-old Dolly received worldwide attention.

Dolly made the newspapers again, about 14 months later, when she answered another clone-related question: Could this technique produce fertile individuals? In May 1998, the Roslin Institute verified that Dolly had given birth to the lamb Bonnie.

Clones Aplenty

Mice

In July 1998, Ryuzo Yanagimachi and his research group at the University of Hawaii's John A. Burns School of Medicine reported that they had cloned more than four dozen mice. Their work provided proof that Dolly was not the first and last such result. Yanagimachi worked with Teruhiko Wakayama of the University of Tokyo in Japan to clone the mice. Their technique was similar to that used to create Dolly. However, instead of following the procedure used by the PPL-Roslin team and employing an electrical pulse to fuse the donor cell to the recipient cell and to trigger cell development, the Hawaiian team used a needle to inject the donated nucleus into the recipient egg, and strontium, a metallic chemical element, to trigger cell development. The mouse cloning team also went a step further and cloned some of the cloned mice.

Cows

Five months later, cows joined the list. Researchers from Kinki University; the Nara Institute of Science and Technology; and the Ministry of Agriculture, Forestry and Fisheries in Japan cloned 10 embryos from one adult's somatic cells. Eight of the 10 grew into calves, 4 of which died at birth or shortly thereafter of environmental causes, the researchers reported. Regardless, their success rate was an improvement over the 1-in-277 survival rate the PPL-Roslin team reported for the work that led to Dolly.

Dogs?

In 1998, some anonymous wealthy pet owners stepped up to fund research they hoped would eventually clone their family dog, Missy. Mark Westhusin, director of the cloning lab at Texas A&M, took on the two-year "Project Missy-plicity." The lab already had one "first" under its belt: it had produced puppies from a surrogate mother. In the 4 September 1998 issue of *Science,* Westhusin said the project should provide useful insights into pet contraception and into efforts to save other endangered species within the dog family.

References

"Hello, Missy," *Science,* 4 September 1998: 1443.
Kato, Y., et al., "Eight Calves Cloned from Somatic Cells of a Single Adult," *Science,* 11 December 1998: 2095–8.

Pennisi, E., "Cloned Mice Provide Company for Dolly," *Science*, 24 July 1998: 495–7.

Travis, J., "Dolly Had a Little Lamb," *Science News*, 2 May 1998: 278.

——, "Ewe Again? Cloning from Adult DNA," *Science News*, 1 March 1997: 132.

——, "A Fantastical Experiment," *Science News*, 5 April 1997: 214.

Wills, C., "A Sheep in Sheep's Clothing?" *Discover*, January 1998: 22–23.

Applications

Practical applications of cloning research abound, particularly as ways to produce pharmaceuticals. By combining genetic engineering and cloning, researchers could create herds of cloned animals or acres of cloned plants that carry a new trait.

The Roslin team has already begun cloning sheep that carry the gene for the human factor IX protein along with a marker gene. Some hemophiliacs take that protein to improve the clotting of their blood. Researchers use well-understood marker genes to track less-familiar genes. These cloned sheep grew from the nuclei of fetal fibroblast cells, instead of from adult cells, as Dolly had. The goal is to produce factor IX and other proteins in the cloned sheep's milk.

Several research groups, including the Japanese team that produced the cloned cows mentioned previously, are also studying how they can clone the animals to produce better-tasting beef. A 1998 Associated Press article reported that cloned vegetables, flowers, and goldfish were already being sold in Japan, and beef from cloned cows had just become available.

References

Pennisi, E., "The Lamb That Roared," *Science*, 19 December 1997: 2038–39.

Wills, C., "A Sheep in Sheep's Clothing?" *Discover*, January 1998: 22–23.

Yamaguchi, M., "Japanese Add Meat from Cloned Cows to Country's Dinner Tables," Associated Press, 8 November 1998.

STEM CELLS

Paramount to development from egg to adult, totipotency was a key to the cloning experiment that resulted in Dolly. On the heels of the introduction of the cloned sheep, another research group announced that it had isolated human embryonic stem cells. These stem cells are derived from totipotent cells of the early embryo and have the ability to differentiate into essentially any type of human cell. From these cells, then, springs forth all of the myriad cells of a human being.

John D. Gearhart of Johns Hopkins Medical Institutions in Maryland made the announcement about the cells at the July 1998 meeting of the International Congress of Developmental Biology. His group isolated the cells from embryonic germ cells (which eventually become sperm and eggs). A few months later, a research team led by James A. Thomson of the University of Wisconsin–Madison published a report detailing how it had generated human embryonic stem cell lines (cells that can be propagated indefinitely) from the blastocyst, the 100-cell stage of development. The Thomson group, which had previously isolated embryonic stem cells from monkeys, also went on to demonstrate that these stem cells could, in fact, differentiate and form various tissue types when implanted in mice.

In other work, a team led by neurobiologist Angelo Vescovi of the National Neurological Institute Carlo Besta in Italy, reported that neural stem cells can switch from generating brain cells, as they do normally, to making blood cells. Unlike the embryonic stem cells, which can differentiate into the whole gamut of cells, neural and other types of stem cells have more limited potential. The researchers conducted their work on mice, introducing the neural stem cells into animals that lacked bone marrow, which produces blood cells. The stem cells generated blood cells.

Potential Applications

The potential applications for human stem cells are manifold. Based on the findings of the neural stem cell work just described, researchers are now considering whether these cells might be a better laboratory-grown source of blood than the more temperamental bone marrow stem cells. Other stem cells might offer similar options. For example, during a U.S. congressional hearing on stem-cell research in 1998, biologist John D. Gearhart predicted that researchers within the next two decades would be able to grow neurons to replace those lost due to Parkinson's disease.

In addition, since embryonic stem cells can differentiate into any type of cell, scientists may one day be able to direct their differentiation and perhaps grow replacement parts for persons with neurodegenerative disorders, or for those who have experienced heart attacks or spinal cord injuries. If the patient's own stem cells were used to grow into the replacement organ, perhaps using some of the cloning techniques described previously, the process would also circumvent the problems associated with organ rejection. While organs made to order may be a distant expectation, they are no longer unimaginable.

Another possibility is the creation of genetically engineered stem cells that could be universal donors for tissue and organ transplants. To create cells like these, researchers would have to remove the numerous proteins that trigger the body's immune reaction. Again, while a great deal of further work is needed, the prospect exists.

Basic Science

Before practical applications become possible, however, scientists must first understand the underlying principles behind the embryonic stem cells' ability to differentiate into all of the different types of cells. Geneticists want to know what genes are involved, and when. Developmental biologists seek to understand how the ultimate paths of a stem cell are determined. A basic understanding of stem cells and their mechanisms will go a long way toward improving the likelihood of success for these highly sought-after applications.

References

Gearhart, J., "New Potential for Human Embryonic Stem Cells," *Science*, 6 November 1998: 1061–2.

Marshall, E., "Use of Stem Cells Still Legally Murky, but Hearing Offers Hope," *Science*, 11 December 1998: 1962–3.

Pedersen, R., "Embryonic Stem Cells for Medicine," *Scientific American*, April 1999: 69–73.

Strauss, E., "Brain Stem Cells Show Their Potential," *Science*, 22 January 1999: 471.

Thomson, J., et al., "Embryonic Stem Cell Lines Derived from Human Blastocysts," *Science*, 6 November 1998: 1145–7.

Travis, J., "Human Embryonic Stem Cells Found?" *Science News*, 19 July 1997: 36.

BIOLOGICAL CLOCKS

Plants and animals keep time. Their biological clocks regulate daily activity, hormone cycles, and other cycles. The clock helps drive a diurnal animal to become active during the day, or a nocturnal animal to liven up after the sun sets. The timings are so apparent that they have crept into the common names of plants and animals. For example, four o'clocks are a type of plant with flowers that open in the afternoon, and common nighthawks are birds that tend to fly after dark.

Our understanding of the biological clock soared in the late 1990s as several research teams began to unravel its secrets and found similarities even across the kingdom divide between plants and animals.

Clock Genes

In 1997–98, researchers distinguished the first mammalian clock genes, *Clock* and *per,* which produce the proteins called CLOCK and PER, respectively, and a third and fourth gene that produce the proteins TIM (short for timeless) and DOUBLETIME. With this revealing view of the genes involved, scientists were able to piece together how the four work in concert to run the biological clock.

Fruit fly researchers demonstrated that morning for the flies arrives when CLOCK binds to another protein and switches on PER and TIM. DOUBLETIME's job is to delay PER and TIM from switching off their genes too early. Eventually PER and TIM override DOUBLETIME, turn off their genes, and wait for CLOCK to "awaken" them the next morning. Another research team showed the same gene-protein loop, and the same or similar genes in mice. Perhaps most surprising, other groups reported that although plants and cyanobacteria use different genes in their clocks, the mechanism matches that of the animals.

Researchers also learned that light fine-tunes the timekeeping ability of the clock in fruit flies by controlling the level of TIM.

In November 1998, three research teams collectively reported that they had possibly identified the molecule responsible for transferring external light to the internal clock in plants, in mice, and in fruit flies. They each pointed to the protein cryptochrome. The finding was another cross-kingdom link. Work published in 1999, however, acknowledged that although cryptochrome has a role in biological clocks—and perhaps is even a component of the clock—it is not the elusive photoreceptor. Additional studies discounted another likely candidate, the light-capturing and vision-enabling molecules in the retina's rod and cone cells. The search continues.

In one twist to the clock story, researchers Scott S. Campbell and Patricia J. Murphy of Cornell University Medical College in New York found that they were able to reset a person's biological clock, not only by exposing the individual's eyes to light, but also by shining light onto the backs of the person's knees. The researchers chose this area of the body because they believe the effects of the light are transferred to the brain by blood, and blood vessels are close to the surface at the backs of the knees. Their finding might be useful for adjusting an individual's biological clock as the person is sleeping.

While scientists find biological clocks interesting from a basic-science point of view, an understanding of the clocks would have practical applications, as well. If researchers were able to determine exactly how biological clocks work, they might be able to prevent jet lag, the period when a traveler's biological clock readjusts to a new time zone, or help workers switch back and forth more easily between day and night shifts.

References

Barinaga, M., "Clock Photoreceptor Shared by Plants and Animals," *Science*, 27 November 1998: 1628–30.

———, "The Clock Plot Thickens," *Science*, 16 April 1999: 421–2.

"First Runner-Up: A Remarkable Year for Clocks," *Science*, 18 December 1998: 2157.

ANTIBIOTIC RESISTANCE

Some strains of bacteria are becoming increasingly resistant to the antibiotic arsenal currently in place to fight them. Health professionals are worried, and biologists and other scientists are trying to find new weapons against the so-called "superbugs."

Survival Tactics

The bacteria's resistance is tied to the rising use of antibiotics. While antibiotics have done a good job of killing disease-causing bacteria, a few individual bacteria inevitably have mutations that allow them to survive the assault. Of those that survive, some may be able to reproduce successfully and pass on the resistance to subsequent generations. This pathway is common to evolutionary processes.

As antibiotic use has skyrocketed, the occurrence of antibiotic-resistant bacteria has become more commonplace. In 1997, for example, patients from three separate locations each contracted an infection of *Staphylococcus aureus* that was resistant to vancomycin, the antibiotic previously used against that infection. This resistant strain of *S. aureus* was still susceptible to other drugs, so the patients survived. Other strains, however, have already proved to be resistant every other drug except vancomycin, so the news of a vancomycin-resistant strain was met with great concern.

These encounters with superbugs aren't unique. Three other species of bacteria have already developed resistance to all known antibiotics. Health professionals are now reporting deaths from previously treatable infections.

New Offenses

While various organizations are suggesting that antibiotics be dispensed only when necessary to lessen the development of resistance, scientists are trying to learn as much as possible about the bacteria and to develop new antibiotics to stop them.

In 1998, for instance, a research team announced that it had identified compounds that prevented the growth of several bacterial strains, including *S. aureus*. Some of these strains had already shown resistance to penicillin and vancomycin. In this research, James A. Hoch of the Scripps Research Institute in California and his research team worked with scientists from the R. W. Johnson Pharmaceutical Research Institute in New Jersey. They were able to derail a two-protein signal system the bacteria use to interact with the outside environment by inhibiting one of the two proteins, called bacterial kinase KinA.

Another report in 1999 illustrated how a research team, led by Min Ge of Princeton University, beefed up vancomycin's effectiveness by adding sugar to specific sites on the antibiotic's chemical structure.

Scientists are continuing the search for better antibiotics to counter the looming threat of the superbugs.

References

Ge, M., et al., "Vancomycin Derivatives That Inhibit Peptidoglycan Biosynthesis without Binding D-Ala-D-Ala," *Science,* 16 April 1999: 507–11.

Levy, S., "The Challenge of Antibiotic Resistance," *Scientific American*, March 1998: 46–53.

Travis, J., "New Antibiotic Dulls Bacterial Senses," *Science News,* 2 May 1998: 276.

Walsh, C., "Deconstructing Vancomycin," *Science,* 16 April 1999: 442–3.

OTHER RECENT FINDINGS

In virtually every topic within the fields of molecular biology and genetics, research in the past few years has surged. These fields intersect with so many other areas both within and outside of the biological sciences, the possibilities for research seem virtually limitless. This section will include a look at a small sample of other active research areas.

Single Genes

As the rush to sequence the human genome continued, many scientists were content with the identification of the genes responsible for various human conditions. For several years, *Discover* magazine has been running an annual listing of new gene discoveries. In the four-year period encompassing the years 1995–98, the magazine's lists included the genes (or gene mutations) believed to be connected with:

- galactosemis, which results in cataracts
- partial epilepsy, a seizure-causing disorder
- Alzheimer's disease
- genetically based and estrogen-prompted breast cancer
- the facial disorder Treacher Collins syndrome
- blindness brought about by retinitis pigmentosa
- the most common skin cancer, known as basal cell carcinoma
- the tremor-inducing Parkinson's Disease
- juvenile-onset glaucoma
- one form of baldness
- a type of heart fibrillation called idiopathic ventricular fibrillation

In some cases, physicians can use this information to screen a suspected carrier of a particular gene and recommend preventive treatments. In the 1990s, for example, some women who were at risk for hereditary breast cancer considered undergoing mastectomies to reduce the probability of facing that disease.

In other work announced in 1999, a group of researchers and medical doctors from Wayne State University in Michigan and Cornell University in New York were able to use genetic analysis to help a sickle-cell-carrying couple give birth to disease-free children. Each parent carried one recessive allele (or gene form) for sickle cell anemia and one normal allele. Persons with two recessive alleles have the disease, which produces sickle-shaped red blood cells. The malformed blood cells lead to circulatory problems and death. The couple participating in the study each carried one normal allele, so any child they conceived would have a 25 percent chance of having two normal alleles, a 50 percent chance of carrying one normal and one recessive allele, and a 25 percent chance of having the two recessive alleles.

The process began with an *in vitro* fertilization, in which the mother's eggs were fertilized outside her body and then implanted. Before the eggs were implanted, however, the research team was able to test them for the sickle cell allele using a procedure called pre-implantation genetic diagnosis. Developed by team member Mark Hughes of Wayne State University, the technique begins by mixing reactive agents with the DNA of a cell that has been extracted from the primitive eight-cell embryo. After amplifying the gene in question into millions of copies, genetic testing determines whether the cell carries the genetic mutation. The ultimate result of the work was healthy, disease-free, twin girls.

The technique has ramifications for parents predisposed to genetic diseases and their children. Formerly, parents had to wait at least nine weeks into a pregnancy for the results of amniocentesis or chorionic villus sampling to determine whether the allele was present in the fetus. With pre-implantation genetic diagnosis, a determination can be made before the fertilized egg is ever implanted.

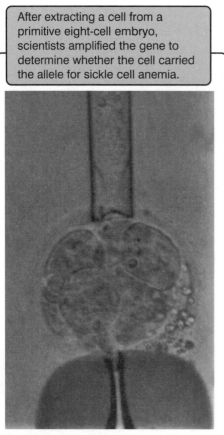

After extracting a cell from a primitive eight-cell embryo, scientists amplified the gene to determine whether the cell carried the allele for sickle cell anemia.

Photo by Dr. Dmitri Dozortsev

As new techniques are developed and additional research is completed, scientists and physicians will begin to narrow further the gap between gene identification and disease treatment, cure, or prevention.

References

Glausiusz, J., "The Genes of 1995," *Discover*, January 1996: 36.
———, "The Genes of 1996," *Discover*, January 1997: 36.
———, "The Genes of 1997," *Discover*, January 1998: 38.
———, "The Genes of 1998," *Discover*, January 1999: 33.
Wayne State University, "Babies Born without Sickle Cell Anemia, Thanks to Genetic Diagnosis Procedure Developed at Wayne State University," press release, 11 May 1999.
Xu, K., et al., "First Unaffected Pregnancy Using Preimplantation Genetic Diagnosis for Sickle Cell Anemia," *Journal of the American Medical Association*, 12 May 1999: 1701–6.

Animal Development

The quest to understand how an organism develops from egg to adult is a dynamic one. In the 1990s, scientists made important discoveries that brought added excitement to the search for knowledge.

Master Control Genes

The field of developmental biology received a big push in 1995 when researchers identified a master control gene in fruit flies. Called *eyeless*, this gene served as the on-off switch for making eyes. A gene is turned on, or "expressed," when its genetic sequence is read through a process called transcription. For transcription to occur, very specific proteins, called transcription factors, bind to the gene and allow the gene's sequence—its instructions—to be read. The gene's product, a protein, is then produced. In the eyeless work, when researchers turned on the gene in different parts of a fruit fly's body, the insect sprouted eyes in the odd locations. They also showed that fruit flies and mammals share the same gene when they implanted a mouse's eye-creating gene in the fruit flies and got eye formation.

Similar work by other labs—including an extensive survey completed by Grace Panganiban of the University of Wisconsin and Steven Irvine of the University of Chicago—showed that invertebrates and vertebrates share a very similar gene, apparently a master control gene, that triggers limb formation, despite the radically different limbs found in the two groups of animals. Vertebrates have an interior skeleton, whereas invertebrates have an outer exoskeleton.

Additional studies showed similarities for other master control genes, leading to renewed discussions about whether certain features evolved independently in different groups of animals and plants, or whether they evolved once, and organisms that share the feature have a common ancestor. Before the findings supporting a similar-gene basis for eye formation in vertebrates and invertebrates, textbooks routinely explained, for instance, that the compound eyes of insects evolved separately from the eyes of vertebrates.

Genes for Differentiation

A variety of research studies in 1999 gave biologists a more in-depth view into how a specific part of an animal's body develops. Several reports showed that one gene could alter the appearance of the forelimbs and hind legs in mice and chickens. One study was conducted by Juan Carlos Izpisúa Belmonte of the Salk Institute for Biological Studies in California, Michael Rosenfeld of the University of California–San Diego, and colleagues. They showed that genetically engineered mice that lacked the leg-specific gene *Pitx1* had hind legs that looked more like forelimbs. When the researchers expressed *Pitx1*—allowed the gene to exhibit its effects—in the wings of chicks, the wings became more leg-like. Scientists believe the genes somehow alter the influence of growth factors and affect the developing limb.

References
"Eyes Everywhere," *Science,* 22 December 1995: 1903.
Travis, J., "Eye-Opening Gene," *Science News,* 10 May 1997: 288.
Vogel, G., "New Findings Reveal How Legs Take Wing," *Science,* 12 March 1999: 1615–6.
Zimmer, C., "Hidden Unity," *Discover,* January 1998: 46–47.

Plant Development

Instead of studying human embryonic stem cells, plant geneticists are investigating shoot apical meristems. Like embryonic stem cells, the meristem cells can differentiate into the different types of cells found in a mature individual.

At the 18th International Congress of Genetics in September 1998, Elliot Meyerowitz of the California Institute of Technology announced that he and his research team found two genes that set the meristems on the path to differentiation. In particular, they studied *Arabidopsis thaliana,* a well-researched, white-flowered plant. They made a series of plant mutants, some of which were missing one or more of the three types of cells that grow into either the inner, middle, or outer layers of the meristem. They also created mutants that exhibited uncontrolled meristem growth without differentiation. The mutants generated normal meristem tissue, although the meristems were significantly larger than the nonmutant variety. With that finding, the researchers deduced that, although the three cell types were not receiving signals to stop growing and differentiate, they were receiving signals to maintain the proper layering and proportions. The research team cloned two genes from the plant and found that their associated proteins appear to communicate between the meristem layers to keep the growth in sync.

At about the same time that the two genes were identified, a University of Michigan research group, led by Steven Clark, identified two proteins that interact with one of the genes, *CLAVATA1*, that the Caltech team detected. They suspected that the two proteins help to control the meristem's growth.

Once scientists learn more about how plants grow, the information may help others who are trying to improve crop yields and productivity.

References

Barinaga, M., "Key Molecular Signals Identified in Plants," *Science*, 19 March 1999: 1825–6.

Normile, D., "Multiplying Knowledge of Cell Division, Plant Growth," *Science*, 11 September 1998: 1591–3.

AIDS from Chimps

Research published in 1999 provided additional evidence that the AIDS-causing virus jumped from chimpanzees to humans. The primary human version of the virus is HIV-1, or human immunodeficiency virus type 1. The simian version is SIV. Chimps rarely experience adverse effects from SIV.

Earlier work had traced the origin of AIDS in humans to a region of Africa that is home to the chimpanzee subspecies *Pan troglodytes troglodytes*. The 1999 research tied the HIV-1 to the same subspecies through a study of the mitochondrial DNA in four chimpanzees that carried the virus. According to the study by Feng Gao of Tulane University in Louisiana and coworkers, each of three strains of SIVcpz (short for the SIV virus in chimpanzees) transferred from the chimps to humans and brought about three types of HIV-1 in humans: the main type, which has spread worldwide, and two other types that are confined to small geographical areas. To strengthen the evidence, scientists hope to test the virus's prevalence in wild chimpanzee populations. Other work had shown a similar jump from SIVsm (the sooty mangabey virus) in a species of West African monkeys to HIV-2, a less-common strain of the virus.

The research heightened concern among some scientists that cross-species disease leaps might not be limited to the closely related chimps and humans, but might extend to more distantly related animals. If this is the case, xenotransplantation—the transplantation of organs into humans from other animals, such as pigs—will face a large hurdle as scientists try to determine whether any porcine diseases can cross to humans.

References

Cohen, J., "AIDS Virus Traced to Chimp Subspecies," *Science,* 5 February 1999: 772–3.

Gao, F., et al., "Origin of HIV-1 in the Chimpanzee *Pan troglodytes troglodytes, Nature,* 4 February 1999: 436–41.

Weiss, R., and R. Wrangham, "The Origin of HIV-1: From *Pan* to Pandemic," *Nature,* 4 February 1999: 385.

Prions

Prions, described as proteinaceous infectious particles or malformed proteins located in the mammalian brain, are probably at the root of mad cow disease and a fatal variant of Creutzfeldt-Jakob disease (CJD) that infects humans, according to several studies. In the mid-1990s, nearly two dozen people contracted the variant CJD and died after eating beef infected with mad cow disease. The variant CJD riddles the brain with holes and causes dementia, tremors, and eventual death in those infected. Mad cow disease also leads to spongiform encephalopathy, the term for the perforated-brain condition, and death in the animals.

The suspected link between the infected cattle and human deaths sparked a short-lived public fear of eating beef, along with a surge in research. In 1997, two studies noted the similarities between the variant CJD and the bovine spongiform encephalopathy. In one, researchers found similarly devastating results in mice that had received tissue containing either the variant CJD or mad cow disease. In addition, the other research team found identical prions in both populations of mice.

The researchers suggested that a single strain of the odd-shaped prions initiate the CJD and mad cow disease by enlisting other proteins to similarly malform and to attack the brain tissue.

Prion-Free Cows

In an article in the December 1998 issue of *Scientific American,* cloning expert Ian Wilmut explained that the cloning technique could have an impact on the incidence of mad cow disease. It could "produce herds of cattle that lack the prion protein gene," he said, adding that it could also eliminate risks associated with those medicines that rely on products derived from cattle.

References

Seppa, N., "Mad Cow Disease, Human Illness Tied," *Science News,* 4 October 1997: 212.

Wilmut, I., "Cloning for Medicine," *Scientific American,* December 1998: 58–63.

Artificial Chromosomes

Just as scientists were taking apart DNA to determine its constituent parts, others were putting it together to make the first artificial human chromosome.

In April 1997, Huntington F. Willard of Case Western Reserve University School of Medicine in Ohio and a research team watched as a chromosome self-assembled from three types of DNA. The chromosome thrived in cells and successfully replicated many times.

The major barrier to making a human artificial chromosome, or HAC, was the lack of understanding surrounding the formation of chromosomes. For example, scientists had little information about the centromere, which is important during cell division when the chromosome's paired chromatids separate and migrate to either of the two resulting daughter cells. Scientists felt they had to know much more about the centromere before anyone would be able to make a chromosome. The Willard group, however, ignored the lack of knowledge about the intricacies of chromosomal formation and, instead, simply added carefully chosen ingredients to cultured human cells and watched. Their ingredients included three types of DNA: normal DNA that contained genes; repetitive DNA sequences believed important for the centromere to function; and DNA from the chromosome end caps, or telomeres. After numerous attempts, a chromosome formed from the three added DNA types.

Artificial chromosomes hold promise as carriers of selected genes into human cells. Current methods using viruses to deliver new genes have limitations—such as gene size or insufficient protein production—that a human artificial chromosome might overcome. Although the Willard group's HAC was less than 20 percent of the length of a normal human chromosome, it was still longer than a viral chromosome and long enough to contain the DNA for genes. As with many other areas of genetics and molecular biology, the research is just beginning.

References

Kher, U., "A Man-Made Chromosome," *Discover,* January 1998: 40–41.
Travis, J., "Human Artificial Chromosome Created," *Science News,* 5 April 1997: 204.

CHAPTER SIX
Social Issues in Biology

O ver the past several years, many areas of biology have become steeped in controversy, particularly as new techniques and technologies give scientists the ability to peer into the genome, manipulate genes, and generate clones from adult cells. This chapter has four sections. The first, on genetics, will include the discussions surrounding cloning, stem-cell research, and the privacy of a person's genetic information. The second section will provide an overview of the debate between those who see transgenic plants as beneficial and those who see them as a menace. The next two sections consider international biodiversity efforts and the recent events surrounding evolution's place in the classroom.

GENETICS

Chapter 5 presented a sample of the many research accomplishments in molecular biology and genetics in the 1990s. Along with those accomplishments, however, have come concerns about how these advances might ultimately be used.

Cloning and Stem Cell Research

When Dolly made her entrance onto the world's stage in 1997 as the world's first mammal cloned from an adult cell, the ethical questions began swirling. The major emphasis was on the potential for human cloning. Fears of cloned dictators creating havoc, and of people creating their clones as sources of spare parts, spread like wildfire. Those fears were fanned in 1998 when the Kyunghee University Hospital in South Korea reported that it had cloned a woman and allowed the resulting cell to divide two times before destroying it. Although other researchers declared that the experiment proved nothing, since the cell was destroyed too early to tell whether the procedure had actually worked, the announcement nonetheless renewed the controversy over human cloning.

In June 1997, U.S. President Bill Clinton followed the lead of the British Parliament, which effectively banned human cloning through its 1990 Human Fertilisation and Embryology Act, and proposed similar action to outlaw the practice. He remarked that the cloning technique "could lead to misguided and malevolent attempts to select certain traits, even to create certain kinds of children—to make our children objects rather than cherished individuals."

Stem cell research evoked similar concerns. The controversy arose when researchers in 1998 announced that they had isolated human embryonic stem cells. The cells have the ability to differentiate into essentially any type of human cell. In 1999, a ban on the use of stem cells obtained from human embryos and fetuses was proposed in the U.S., followed by a letter to President Clinton supporting stem cell research, signed by 33 Nobel laureates (see Chapter 8). As of 1999, the United States, Great Britain, France, Germany, Norway, and other countries have ceased public funding of such research and/or have imposed other restrictions on the use of embryonic and fetal tissues. The issue divides those who fear an abuse of the technology and those who appreciate its potential to treat illnesses.

References

"Immortal Cells Spawn Ethical Concerns," *Science*, 18 December 1998: 2161.
Travis, J., "More Cloning News Closed Out 1998," *Science News*, 16 January 1999: 43.
Wills, C., "A Sheep in Sheep's Clothing?" *Discover*, January 1998: 22–23.

Privacy Issues

A heated debate also took shape in 1998 over a proposed genetic databank in Iceland. After months of wrangling, the Icelandic parliament voted to allow the deCODE Genetics company of Iceland to set up and use a nationwide database of the citizens' genetic information. Under this plan, doctors would enter their patients' health information onto deCODE-provided terminals installed in all of the country's health care facilities. The data would then be compiled into a national database that includes individual records that are listed by code instead of patient names. Under the agreement, the company has exclusive rights for 12 years to market that data.

The company hopes to take advantage of the database and the country's unusually small gene pool to detect disease-related genes and develop treatments, and it has already inked a deal with the Hoffmann-La Roche pharmaceutical company, which will develop and market drugs to treat genetic disorders whose genes are discovered via the database. The Icelandic government believes the database will improve the health care system in Iceland and boost the economy with both dollars and technology-related jobs. Opponents of the agreement, however, believe it violates the privacy of citizens, who have no say in whether their records are added to the database. Although individuals' names do not accompany the records, some opponents fear that sensitive personal information could become available, further violating citizens' privacy. Their concerns heightened when an addition to the bill presented the option for the company to connect the database to citizens' genealogical records. Based on a poll the company had taken, however, deCODE representatives said the public supported the database by a margin of four to one. Many of the detractors, they added, were from competing companies. Patient groups, physicians, scientists, and ethicists will be watching Iceland over the next few years to see how the story unfolds.

Public Perception

Views about privacy issues are particularly poignant among those who have histories of genetic disorders in their families. To begin to understand their point of view, researchers E. Virginia Lapham and Chahira Kozma of Georgetown University Child Development Center in Washington, D.C., and Joan O. Weiss of the Alliance of Genetic Support Groups in Maryland interviewed 332 members of genetic-disorder support groups. The support group members had at least one genetic disorder in their families. In this survey of their perceptions, many reported apprehension about their genetic records being made available to employers and insurance companies. One in four perceived that life insurance companies had refused them coverage based on their family history, and more than one in five believed the same about health insurance companies. In addition, 13 percent felt they either weren't offered a job or were dismissed from a job because of their genealogies. These anxieties and beliefs led 9 percent those surveyed to forgo genetic tests, and nearly one in five to withhold genetic information from insurers and employers.

Public perception may or may not reflect reality, but it will likely have a profound impact on the future of genetic testing and genetic research.

References

Enserink, M., "Iceland OKs Private Health Databank," *Science*, 1 January 1999: 13.
———, "Physicians Wary of Scheme to Pool Icelanders' Genetic Data," *Science*, 14 August 1998: 890–891.
Lapham, E., et al., "Genetic Discrimination: Perspectives of Consumers," *Science*, 25 October 1996: 621–24.

Genetic Enhancement

Genetic enhancement in humans is a term applied to the improvement or addition of a genetic trait, usually one that goes beyond the treatment or avoidance of a disease. Such changes in an individual's genetic assemblage present the possibilities—however remote—that parents could choose the color of their baby's eyes, determine how pretty or handsome their offspring would be, or select specific intellectual or physical capabilities that would lay the groundwork for a career as a concert pianist, a neurosurgeon, or a major league sports player. Does it follow that genetic studies might lead, however inadvertently, to "designer babies?"

Across the board, scientists and politicians have denounced genetic enhancement in humans. "To raise the athletic capabilities of a schoolyard basketball player to those of a professional or to confer the talents of Chopin on a typical college music professor is the sort of genetic enhancement that many find troublesome," noted Jon W. Gordon, Mathers Professor of Geriatrics at the Center for Laboratory Animal Sciences, Mount Sinai School of Medicine in New York, in an article he wrote for the 26 March 1999 issue of *Science*. While that concern is understandable, it is premature, he said. The ability to make such genetic enhancements is "not practicable in the near future." He urged that basic genetic research be allowed to continue, without legislative restrictions, in the hopes that it could lead to beneficial medical treatments or to new insights into life processes. Legislation designed to curb irresponsible uses of the technology would likely also have the effect of restricting current work, which might quell potentially beneficial findings and applications.

The ability to make complicated genetic enhancements in humans may not become a reality anytime soon, but scientists in September 1999 announced that they had been successful in enhancing memory and learning abilities in mice by adding one gene that improves communication between brain cells. Joe Z. Tsien of Princeton University and his colleagues, who published their results in *Nature,* added, "Our results suggest that genetic enhancement of mental and cognitive abilities such as intelligence and memory in mammals is feasible." In the 4 September 1999 issue of *Science News,* Princeton colleague Lee M. Silver, who investigates the social aspects of genetic engineering, remarked that Tsien's study "demolishes the argument of those who claim that things like memory, learning, and intelligence are so complicated that scientists will never be able to figure out ways to enhance those traits." He added that the potential to create smarter humans could be as little as 25 years away.

According to many researchers, they are already receiving requests from individuals seeking to take advantage of genes that have been found to confer certain traits. An October 1997 article in the *Washington Post* described such requests that genetics researchers have received from medical doctors: one was a request for genetic enhancement to change the color of a patient's skin and another was to generate bigger muscles in athletes. The article also noted that polls of Americans in 1986 and 1992 indicated that 40 to 45 percent had no problem with genetically altering a person's physical or mental endowment. In the article, gene therapist W. French Anderson of the University of Southern California remarked, "Genetic enhancement is going to happen. Congress is not going to pass a law keeping you from curing baldness."

Both research advancements and surveys favoring genetic enhancement have fanned the controversy over genetic enhancement. Many scientists and others envision a society that pits those who have had genetic enhancements, which would likely be the wealthier citizens who could afford them, against those without the alterations. Science-fiction movies and authors have expanded upon those fears to incorporate testing of a person's genetic component in everyday life, and giving those with the favored enhancements the better options in careers, social status, or other areas.

In a paper published in the *Journal of Medicine and Philosophy,* William Gardner of the department of psychiatry at the University of Pittsburgh wrote that once developed, worldwide and complete containment of genetic-enhancement technologies will be unfeasible. At the beginning of the paper, he asked, "In other words, if it becomes possible to produce genetically enhanced children with attributes desired by parents, can we successfully prohibit parents from doing so?" He concluded the lengthy article with the statement, "[if] the argument presented here is correct, then prohibiting genetic enhancement would be similar to, but perhaps even more challenging than the (so-called) control of nuclear weapons."

This debate will continue, and it promises to become more heated as technology advances.

References

Gardner, W., "Can Human Genetic Enhancement Be Prohibited?" *Journal of Medicine and Philosophy* 20 (1995): 65–84.

Gordon, J., "Genetic Enhancement in Humans," *Science*, 26 March 1999: 2023–4.

Travis, J., "Gene Tinkering Makes Memorable Mice," *Science News*, 4 September, 1999: 149.

Weiss, R., "Cosmetic Gene Therapy's Thorny Traits," *Washington Post*, 12 October 1997: A01.

Patent Issues

The desire to sequence the human genome has created a race driven by patents. As described in Chapter 5, many academic researchers feared that a private company would beat the Human Genome Project's team and patent the human genome. At the head of the race were two groups: J. Craig Venter and his Celera Genomics company in Maryland, and the Human Genome Project with primary research teams at Washington University in Missouri, Baylor College of Medicine in Texas, Whitehead Institute for Biomedical Research in Massachusetts, and the Sanger Centre in England.

While a private company cannot patent the genome itself, it can patent the process it used to generate the genome and any information so gleaned from that process. Therefore, if Celera succeeds in sequencing the genome using a particular process, it would own the data it collected. Others would have the option to obtain that data through Celera, or develop their own process and collect the data themselves. The latter option is impractical for most researchers, who have neither the funding nor the resources to take on such an enormous task.

U.S. law awards patents to inventions that are novel, are not obvious, and have utility. The U.S. Patent Office determined that human DNA fragments fit all three requirements, and the office had issued more than 1,500 patents by early 1999. Actually, the patents cover the processes used to generate the DNA sequences, along with the information gained from that process—the DNA fragments.

Many scientists believe DNA should not be patentable, and instead should remain in the public domain.

References

Pennisi, E., "Academic Sequencers Challenge Celera in a Spring to the Finish," *Science*, 19 March 1999: 1822–3.

Wertz, D., "Patenting DNA: A Primer," *Gene Letter*, February 1999: <http://www.genesage.com/professionals/geneletter/archives/dna1.html>, accessed 30 March 2000.

Many Questions

As researchers, ethicists, and politicians wend their way through the issues surrounding genetics, they are facing many questions.

One that touches on the Iceland situation is the ownership of a person's genetic information. Does is belong to the individual, the doctor, the medical community, the government, or a private company? Should people be able to keep their genetic history to themselves? What if that information could help identify disease genes and possibly develop treatments? Should an individual's right to privacy outweigh potential benefits to the community's overall health? Can both be served?

As genetic testing increases, physicians will also be facing ethical issues. Should they share the results of genetic testing with their patients only if a discovered illness is treatable? What if the results portend an untreatable, fatal illness? If genetic testing of a fetus picks up something other than a fatal disease—perhaps a potential disability, such as blindness—should the parent be informed? How about color blindness? What if the physician suspects that the parent will seek to abort a fetus with a disability, no matter how minor?

Future questions may also shift to whether parents should have any control over the traits of their offspring, and if so, the amount of that control. These questions inevitably lead back to the potential for genetic enhancement.

As genetic research continues and new findings mount, the ethical and social questions will increase in number and significance.

References

Strauss, E., "The Tissue Issue: Losing Oneself to Science?" *Science News*, 20 September 1997: 190.

TRANSGENIC PLANTS

Farmers are growing transgenic (genetically engineered) plants that stave off damaging pests, have increased harvests, or produce enhanced vegetables. The U.S. had approved around three dozen such crops by the beginning of 1999, and more than a quarter of the soybean seeds sown on the nation's farms were genetically modified varieties. Notwithstanding, some scientists and other citizens are recommending limitations in their use, and a number even propose bans. Europeans have taken a particularly firm stance against genetically modified crops.

Scientists' Concerns

Some scientists worry that transgenic plants might cause environmental problems if their newly added genes jump species, or if the enhanced pest-fighting abilities lead to the creation of "superpests," much as the overuse of antibiotics has generated "superbugs" (see the section on antibiotic resistance in Chapter 5).

The bacterium *Bacillus thuringiensis*, also known as Bt, is a popular addition to transgenic crops. Corn yields can falter if a crop is infested by any of a number of corn borers, which attack the cobs and the stalks. Bt kills nearly all of these pests, so Bt corn is economically attractive to farmers. Concern over its use arises because Bt excels in this role. Like antibiotics that wipe out all but the most resistant bacterial strains, Bt kills all but a few resistant pests, a situation that may then lead to Bt-resistant insects becoming the predominant variety. The artificial overdose of the bacterium, which has been described as nature's most important pesticide, may one day be obsolete against the new, Bt-resistant insect populations.

Before transgenic plants became available, farmers periodically sprayed Bt on their crops. The time between sprayings was sufficient for the nonresistant corn borers to repopulate. Opponents to the overuse of Bt recommend that farmers refrain from planting exclusively the modified corn. The untreated part of the crop would give nonresistant insects a haven to reproduce, and would prevent the resistant insects from overrunning the normal corn borer population.

Legal Wrangling

The increased use of Bt plants and the growing threat of resistant crop-damaging insects are of particular worry to organic farmers. The International Federation of Organic Agricultural Movements, a global federation of organic farmers, processors, and certifiers, joined forces with Greenpeace International in February 1999 to file a lawsuit against the U.S. Environmental Protection Agency (EPA) for permitting the marketing of Bt plants, which they felt would eventually render the *Bacillus thuringiensis* bacterium useless against pests that had become resistant. Additional plaintiffs in the lawsuit included the Center for Food Safety, other environmental organizations, and numerous individuals.

Citing their belief that the widespread planting of Bt plants would eventually make Bt useless as a pest-control measure, the suit charged the EPA with "the wanton destruction" of Bt. The suit described Bt as having natural toxins that "are essential to a 21st century agriculture based on biological controls and not the use of synthetic insecticides." It further claims that the EPA violated acts that prohibit that agency from registering pesticides that cause unreasonable adverse environmental effects, and require environmental impact statements on major federal actions. The suit states, "The EPA has failed to prepare a programmatic environmental impact statement analyzing the environmental, socio-economic and cumulative impacts of its program registering genetically engineering Bt plants."

European Rejection

A 1998 poll in the United Kingdom reported that more than three-quarters of those surveyed favored a ban on transgenic crops. Other European nations have banned various genetically engineered plants from their farm fields. The rejection of these modified plants stems in part from fears that the transgenic crops could affect wildlife areas that, in Europe, are commonly intermingled with farm fields.

Unlike the United States, Europe's farms are mainly small, family-run ventures. Many of them are adjacent to open fields or forests that are home to wildlife. A spokesperson for the Royal Society for Protection of Birds, a British conservation charity, remarked in the 7 August 1998 issue of *Science*, "Narrow strips of land around field margins left to grow weeds and other wild plants provide a vital habitat and food source for many creatures, and are highly vulnerable to changes in management practices." Others complain that studies of the effects of transgenic plants on wildlife are severely lacking or nonexistent. Both the Royal Society for Protection of Birds and English Nature, a conservation organization, sought a governmental ban on the crops until such research was conducted.

New Research

The controversy continued when a group of U.K. researchers reported in March 1999 that ladybird beetles, or ladybugs, were adversely affected after eating aphids targeted by genetically modified potatoes. The potatoes contained a gene to deter the aphids without killing them. Ladybird beetles that fed on these aphids experienced a considerable drop in egg viability. The research group included A. Nicholas E. Birch and Irene E. Geoghegan of the Scottish Crop Research Institute and their colleagues from the University of Cambridge, the Biomathematics and Statistics Scotland, Dundee Unit, and the University of Durham.

In May 1999, another group of researchers found that the pollen from Bt corn was deadly to monarch butterfly (*Danaus plexippus*) larvae. Led by entomologist John E. Losey of Cornell University in New York, the researchers reported that the Bt corn's pollen led to death or slow growth in the monarch larvae. Although monarch butterfly larvae don't eat corn, they do inadvertently ingest the pollen when it lands on milkweed (*Asclepias curassavica*), the only plant in their diet. The potential for contaminated milkweed occurs because milkweed commonly grows near

Monarch butterfly caterpillars *(Danaus plexippus)* can be the unintended victims of transgenic corn.

Photo by Ben Czinski

and sometimes in farm fields of the Midwest, where corn is one of the main crops. The wind can blow corn pollen up to 60 yards from the edges of a farm field. In addition, the Midwest is home to about half the U.S. population of monarchs.

The researchers conducted their experiment by misting water onto milkweed plant leaves, then dusting them with the Bt corn's pollen. They allowed three-day-old monarch caterpillars to feed on the leaves for four days, documenting their survival and weight. At the same time, they recorded each caterpillar's survival and weight following a diet of leaves that had either received a dusting of untransformed pollen or had received no dusting of any pollen. They found that the only mortalities occurred in the larvae that fed on the leaves with the Bt pollen. The researchers also noted that the caterpillars ate fewer of the pollen-treated leaves, and particularly decreased their feeding of the leaves that had received the Bt pollen dusting. With the reduced food intake corresponding to the leaves with Bt pollen, the surviving larvae grew more slowly than their counterparts, with final weights less than half that of the caterpillars that ate pollen-free leaves.

References

Birch, A., "Tri-trophic Interactions Involving Pest Aphids, Predatory 2-Spot Lady-birds and Transgenic Potatoes Expressing Snowdrop Lectin for Aphid Resistance," *Molecular Breeding*, March 1999: 75–83.

Cornell University, "Toxic Pollen from a Widely Planted, Genetically Modified Corn Can Kill Monarch Butterflies, Cornell Study Shows," press release, <http://www.news.cornell.edu/releases/May99/Butterflies.bpf.html>, accessed 2 February 2000.

Greenpeace, et al. v. Browner, U.S. District Count for District of Columbia, filed 18 February 1999.

Losey, J., "Transgenic Pollen Harms Monarch Larvae," *Nature*, 20 May 1999: 214.

Williams, N., "Agricultural Biotech Faces Backlash in Europe," *Science*, 7 August 1998: 768–71.

INTERNATIONAL BIODIVERSITY EFFORTS

About 80 percent of Earth's biodiversity is in countries that share only 6 percent of the world's scientists. Those statistics point up the need for cross-cultural relationships between biologists and local residents. In many biodiversity-rich countries, residents are financially poor and living in a close relationship with the land. Scientists are increasingly finding that the benefits of forming close relationships with indigenous populations can be immeasurable, in terms of learning about plants and animals that were previously unknown or little-known to modern science, and particularly about herbal medicines. Many native peoples are also providing valuable input in the sometimes-conflicting needs of the community residents and of the overall ecosystem.

Community-Based Conservation

Community-based natural resource management is bringing the wants and needs of local residents into the discussions about proposed conservation measures. On the one hand, biologists are striving to maintain or improve ecosystems and protect species. On the other, local farmers are trying to make a living from the same land and are battling many of the same species, which are trampling or eating their crops.

In an article published in the 19 March 1999 *Science*, a group of researchers from the University of California–Berkeley, WWF–World Wildlife Fund, University of Pretoria in South Africa, and other universities and organizations, noted that the success of community-based natural resource management in South Africa hinges on two factors. Local communities, instead of outside authority figures, must manage their wildlife, and the community must reap rewards from that management. One program with a 17-year history of following those mandates has been running in Zimbabwe. Villagers now deal with tourism outfits to set up controlled wildlife-viewing or wildlife-hunting safaris. Some species that were once seen as pests are now viewed as income-generators.

Scientists and policy makers are now considering such community-based, resource management efforts as the best way to preserve endangered habitats and species, particularly in the species-rich nations: While biologists can contribute important scientific information, the local people have a better grasp of the social and economic issues they face, so they are able to hammer out workable arrangements with benefits for themselves and for the environment. The authors of the 19 March 1999 paper concluded, "Scientists who wish to be effective in conserving biodiversity for further generations will have to learn how to operate in this new arena."

Cross-Cultural Agreements

Many scientists and pharmaceutical companies are now leading efforts to ensure that indigenous populations receive a share of any economic gains that arise from their contributions. Such deals between native people and pharmaceutical companies or other organizations fall under the heading "bioprospecting." On the other hand, some scientists and advocates for native people call the give-and-take relationships and the low-percentage returns for local communities little more than thinly veiled public-relations ploys that will have few lasting benefits to the people.

A strong proponent of science-giving-back-to-the-community is Paul Alan Cox of the National Tropical Botanical Garden. During a presentation by Cox at the 1999 meeting of the American Association for the Advancement of Science, he explained how he works very closely with native healers in various areas of the world, living in the villages and learning about their medicines. On one trek, for instance, Cox traveled to a Samoan village to seek out potential herbal medicines. He located a plant-derived drug, now called prostatin, that has anti-HIV activity. While in Samoa, his work was threatened by loggers who were destroying the very plants he was studying. He chronicled his work in the book *Nafanua: Saving the Samoan Rain Forest*. Through one of the many give-and-take relationships he has since developed with various villages, 50 percent of the profits will go to the tribal healer's people. Cox believes that such arrangements help to protect individual rain forest plants by emphasizing the value of the entire ecosystem.

Company-Community Deal

In what many are heralding as a giant step toward environmental coop-eration, the Tropical Botanic Garden and Research Institute of India reached an agreement in 1999 with a native community in the country to market an herbal tonic derived from the arogyapacha plant, *Trichopus zeylanicus*. The native community, the Kani tribe, collects the plants from nearby forests. Institute scientists isolated and tested the plant's active ingredient, and the Aryavaidya Pharmacy Coimbtore Ltd. created the tonic, which purportedly boosts energy and immunity. The native com-munity and the institute will share 2 percent of the net profits from the tonic, and they were scheduled to receive their first combined payment of $21,000 in early 1999.

References

Bagla, P., "Model Indian Deal Generates Payments," *Science,* 12 March 1999: 1614–5.

Cox, P., "Conservation in Jurassic Park: Endangered Plants and the National Tropical Botanical Garden," topical lecture, American Association for the Advancement of Science 1999 annual meeting, Anaheim, CA, 22 January 1999.

————, *Nafanua: Saving the Samoan Rain Forest,* W.H. Freeman and Co., 1997.

Getz, W., et al., "Sustaining Natural and Human Capital: Villagers and Scientists," *Science*, 19 March 1999: 1855–6.

"Systemic Biology for the New Millennium," session at the American Association for the Advancement of Science, Philadelphia, 14 February 1998.

EVOLUTION

Evolution is a fundamental part of the biological sciences. It is a theory, in the scientific sense of the word. In colloquial usage, the word theory refers to an unproven idea, or a guess. For example, a child can have a theory about how Santa Claus visits every house on Christmas Eve, or a mother can have a theory about why her teenager is so moody. In scientific terms, however, a theory is a conclusion based on rigorous evidence. Thus, the theory of evolution is much more than just one of many ideas. It is a fundamental principle based on monumental evidence, including examinations of the fossil record, genetic evidence, and exten-sive comparisons of species similarities and differences. The theory of evolution is supported overwhelmingly by scientists, men and women who are trained to ask questions, design experiments, view and compare data, and—only after comprehensive analyses—to then draw conclu-sions.

Despite the importance of, and evidence for, the theory of evolution, the 1990s saw a conflict again arising over its position in the curricula of primary and secondary schools in the United States. The controversy was led mostly by so-called creationists, who believe that a supreme being placed the complete diversity of fully formed species onto Earth at one time. They also believe that Earth and the universe have not changed substantially over time, thus discounting astronomers' Big Bang Theory, which describes how the universe began; geologists' plate tectonics theory, which explains how Earth's continents have moved over time; and even the methods scientists use to date fossils.

The Challenges

Boards of education in several states became embroiled in the controversy, with some debates spilling into the legal and legislative systems. In 1996, for example, the Tennessee, Georgia, and Ohio legislatures entertained bills that would have allowed the teaching of creationism in addition to evolution in their schools' biology courses. While all three bills eventually failed, scientists nationwide became concerned that some science teachers might simply omit any mention of evolution rather than face potential heat from creationist parents.

Polls have also shown that the public remains unconvinced about many scientific theories. A recent poll revealed that more than half of American adults weren't convinced that humans evolved from earlier life forms (other polls have placed the number at around 45 percent). The same poll, which was conducted by the National Science Foundation, showed that nearly two-thirds didn't believe the Big Bang Theory, and slightly more than half thought that humans and dinosaurs lived on Earth at the same time. In general, scientists felt the results indicated faults in the extent or in the presentation of scientific topics in schools, and some organized to promote better science education.

One such group is the Tennessee Darwin Coalition, whose members suspect that the vast majority of those adults who said they didn't believe the theory of evolution simply were unaware of the vast amount of evidence supporting it. To rectify the situation, the coalition held the first annual Darwin Day in 1997 and opened the event to the public. The coalition invited renowned evolutionary biologist Douglas Futuyma of the State University of New York–Stony Brook as a guest lecturer, and it provided information and ran documentaries about the theory of evolution. Additionally, the group established a Web site at <http://fp/bio.utk.edu/darwin/default.html> to reach a larger audience.

The National Academy of Sciences also issued a report, "Science and Creationism: A View from the National Academy of Sciences," which explained that evolution is grounded in scientific investigation and is testable, while creationism is not. The report added, "No body of beliefs that has its origin in doctrinal material rather than scientific observation, interpretation and experimentation should be admissible as science in any science course."

Alabama and Kansas

In 1995, the Alabama State Board of Education caused a nationwide commotion when it required all of its state-approved biology textbooks to carry a disclaimer, which begins,

> This textbook discusses evolution, a controversial theory some scientists present as a scientific explanation for the origin of living things, such as plants, animals and humans.
> No one was present when life first appeared on Earth. Therefore, any statement about life's origins should be considered as theory, not fact.

Scientists, pro-evolution politicians, and many others condemned the action. Many took the board of education to task for misleading statements in the disclaimer. The first sentence, for example, gives the impression that a small number of scientists support the theory of evolution, when actually the vast majority stand firmly behind the theory. The third sentence appears to use the term "theory" in its colloquial sense, meaning "just one of many ideas," rather than in its scientific sense, meaning "a conclusion based on substantial and abundant scientific evidence."

The disclaimer also notes that microevolution and macroevolution are two very different things, with the former something that "can be observed and described as fact," and the latter an "unproven belief." Basically, microevolution describes genetic changes within species. Different characteristics among dog breeds fall under the heading microevolution, as does the color change in the peppered moth, *Biston betularia*, from mostly pale gray to black, which provided camouflage by more closely matching the newly soot-covered trees that resulted from the Industrial Revolution. Macroevolution refers to numerous minor changes that accumulate over long periods of time and eventually change a population or an entire species to such a degree that it evolves into a new species. The creationists' view, along with the disclaimer, acknowledges that species can change a little bit over a short period of time, but not a lot over a long period of time. Scientists, on the other hand, view evolution as a theory that embraces the full range of species changes over time.

The evolution debate again heated up in 1999 when the Kansas Board of Education voted to eliminate evolution, along with mention of Earth's geological age and associated theories, from the state's standardized tests. While the decision fell short of banning the teaching of evolution (court rulings had previously prohibited straightforward bans), it gave teachers a rationale for removing evolution from their curricula. The decision drew fire from scientists and from college and university administrators across the nation. Even Kansas governor Bill Graves called the decision "terrible, tragic, [and] embarrassing." Some scientists recommended that universities and colleges deny admission to students trained in the Kansas school system, and some suggested additional entrance exams for Kansas students to ensure that they understood the theory of evolution and its wide-ranging implications. Others felt more intense educational campaigns should begin to explain evolution to not only Kansas students, but the public in general.

Despite scientists' overall denunciation of attempts to discredit the theory of evolution, the Kansas anti-evolution decision was not the last. Other local and state school boards continue to consider similar actions, and the scientific community likewise perseveres in its search for individual and possibly sweeping solutions to this issue.

References

Chasan, R., "Fighting Back for Science," *BioScience*, January 1998: 8.

"Creation v. Evolution: Counter-attack," *The Economist*, 17 August 1996: 26–27.

Dawkins, R., "The 'Alabama Insert': A Study in Ignorance and Dishonesty," Franklin Lectures in Science and Humanities, Auburn University, 1 April 1996, *Journal of the Alabama Academy of Science* 8 (1), January 1997.

Holden, C., "Kansas Dumps Darwin, Raises Alarm across the United States," *Science*, 20 August 1999: 1186–7.

Schmidt, K., "Creationists Evolve New Strategy," *Science*, 26 July 1996: 420.

"Science and Creationism: A View from the National Academy of Sciences," second edition, National Academy of Sciences, Washington, DC: National Academy Press, 1999. Also available at <http://bob.nap.edu/readingroom/books/creationism/index.html>, accessed 2 February 2000.

CHAPTER SEVEN
Biographical Sketches

T his chapter highlights some of the researchers mentioned elsewhere in this book. Most of the biographies are of people who
were educated as biologists. A few of them, however, were trained
in fields like chemistry and physics. This range of backgrounds shows the
diversity within the field of biology-related research, and it also suggests
the myriad of opportunities available for interdisciplinary studies.

In addition, the individuals featured here range from doctoral students
to well-known scientists. Both well-seasoned and budding biologists can
easily find areas within the biological sciences that are in need of researchers. The fields covered here fall under an array of headings:
evolution, genetic engineering, the origins of life, plant taxonomy, relationships within ecosystems, and animal predation methods.

Information about other researchers mentioned in this book (but not
included in this chapter) is often available through the World Wide Web
pages of the university or institution with which they are affiliated.

Jean Chmielewski (1961–)

Jean Chmielewski is a chemist working on a grand biological question:
How did life begin? In 1998, she added a new twist to the puzzle by
suggesting that proteins—instead of the previously suspected nucleic
acids, or DNA and RNA—were a lead player in the evolution of life.

Chmielewski's research interests in biology-related topics began when she was an undergraduate chemistry student at St. Joseph's University in Pennsylvania. There, she was part of a lab studying prostaglandin construction. Prostaglandins are substances that stimulate the body in a variety of ways, such as enhancing hormonal effects. She went on to earn a Ph.D. from Columbia University in New York, where she studied substances that mimic the enzymes involved in transferring amino groups from one amino acid to form another amino acid. The building blocks of proteins, amino acids are each composed of a carboxyl group (COOH) and a specific amino group (NH_2) bound to a carbon atom.

Chmielewski then accepted two National Institutes of Health (NIH) postdoctoral fellowships, the first at Rockefeller University in New York and the second at the University of California–Berkeley, before joining Purdue University in Indiana as an assistant professor in 1990. She became an associate professor in 1996.

Chmielewski has also accepted an NIH First Award and a National Science Foundation National Young Investigators Award, and is a fellow of the Alfred Sloan Foundation and the Exxon Education Foundation.

Her research group has a number of ongoing projects, including research into the enzymes of HIV, or human immunodeficiency virus; inhibition of DNA binding, which may play a role in anti-cancer therapeutics; and self-assembling peptides. The paper describing proteins as the possible progenitors of life stemmed from her group's studies of a self-replicating, peptide-based system. The system contains four peptides—the units that constitute proteins—and can replicate itself and also can adapt to environmental changes.

Building upon work done at Scripps Research Institute, Chmielewski's group designed a more complex system that not only produced the four expected proteins from the four peptides, but also seven new copies through a process called cross-replication. In a press release, Chmielewski reported, "This is the most complex replicating system of its kind to date. This work clearly demonstrates that peptides should be considered in discussions over the nature of molecular origins of life."

References

Chmielewski, J., biographical sketch provided 10 June 1999.

Chmielewski, J., et al., "Selective Amplification via Auto- and Cross-Catalysis in a Replicating Peptide System," *Nature,* 3 December 1998: 447–50.

Purdue University, "Purdue Study Breathes New Life into Question of How Life Began," press release, 3 December 1998.

Barbara Ertter (1953–)

Overseeing some 1.8 million plant specimens, Barbara Ertter is collections manager and curator of western North American flora at the

University Herbarium and Jepson Herbarium of the University of California–Berkeley. Besides the monumental tasks associated with those titles, she also serves as chair of the Rare Plant Scientific Advisory Committee of the California Native Plant Society, and attempts to introduce and reinforce taxonomic concepts to the general public and to decision makers. Perhaps the greatest testament to her expertise as a botanist are the requests she receives from other botanists to help with the identification of various plant species.

A fourth-generation Idaho native, Ertter's interest in plants became firmly ingrained when she entered Albertson College of Idaho and joined a botanical expedition to Leslie Gulch in eastern Oregon. En route to earning a bachelor of science degree in biology, she completed a senior honors paper on wetland plants of southwestern Idaho and adjacent Oregon.

While continuing her education at the University of Maryland at College Park, she spent two summers as a biological aide, mapping vegetation in Bennett Hills and southeastern Idaho for the big game range inventory project of the Idaho Fish and Game Department. She also served as a graduate teaching assistant, performed research on a southern California plant, and described two new taxa in the *Oxytheca* genus at the University of Maryland before graduating in 1977 with a master of science degree in botany.

Ertter finished her formal education with a doctorate in biology from the City University of New York in a joint program with The New York Botanical Garden. Her dissertation complemented her description of three new taxa of the *Juncus* genus of dwarf rushes. While in the doctoral program, she was a graduate herbarium fellow at the New York Botanical Garden, where she did a good deal of collecting and species identification work, and was a contract supplier for the United States Department of Agriculture's Economic Botany Laboratory, where she collected plants for anti-cancer screening.

Upon graduation, the botanist spent a short time as a lecturer in the biology department at the University of Texas at Austin before accepting a position in 1982 as the university's herbarium curator. In 1985, she joined University and Jepson Herbaria, and in 1994, she also accepted the title of curator of western North American flora for the herbaria.

Ertter has a particular interest in western plants and has become especially captivated by the Potentilleae of the large Rosaceae family and its pattern of "island biogeography" on the desert mountains. She has also taken up the historical aspects of western floristics, with an emphasis on California resident botanists and on "changing perceptions of taxonomy as a science, including gender issues and species definitions."

In addition to a long list of publications, Ertter is a reviewer for papers in *Brittonia, Great Basin Naturalist, Madroño, Novon, Phytologia, Sida,* and *Systematic Botany*, and she is a regional reviewer for *Flora of North America*. She is also a member of numerous organizations, including the American Institute of Biological Sciences, American Society of Plant Taxonomists, Botanical Society of America, several statewide societies, and The Nature Conservancy.

References

Ertter, B., curriculum vitae, 25 April 1997.
Milius, S., "Unknown Plants under Our Noses: How Much Backyard Botany Remains to Be Discovered?" *Science News*, 2 January 1999: 8–10.

Brian D. Farrell (1955–)

Can it be shown that plants and animals co-evolve? That is, is it possible to demonstrate that plants and animals play off of one another and adapt over time?

To study this far-reaching and complexity-ridden question, Brian D. Farrell didn't limit his study to one plant species and one animal species, or even to a few species of each. Instead, he took on the diverse flowering plants known as angiosperms, and the beetles, or Coleoptera, the order of animals with by far the largest number of known species on Earth. In particular, he investigated the plant-eating beetles.

Scientists had previously suspected the flowering plants and the Coleoptera co-evolved, but new reports had begun to question whether the relationship between the two was as strong as suggested. After all, beetles date back 250 million years, while angiosperms had their start only 100 million years ago.

Farrell set out to investigate the association by comparing 115 DNA sequences and 212 morphological characters of members of the Phytophaga, the largest grouping of the herbivorous beetles, and then reconstructing the grouping's phylogeny. He overlaid the reconstructed phylogeny on that of the angiosperms and showed that new angiosperm-feeding lineages of beetles repeatedly correlated with overall great diversifications of the beetles. In other words, as plant-eating beetles began eating angiosperms, the number of new species grew by leaps and bounds. The angiosperms opened the door for the adaptive radiation of the beetles. In "Inordinate Fondness," he concluded, "The success of the order Coleoptera thus seems to have been enabled by the rise of flowering plants."

Farrell earned a bachelor of arts degree in life sciences, with concentrations in botany and zoology, from the University of Vermont, and a master's degree and Ph.D. from the University of Maryland. During his graduate education, he served as graduate teaching fellow in the entomology department at the university and curatorial fellow in the entomology department at the Smithsonian Institution. He completed his dissertation on phylogenetic studies of interactions between plants and insects.

Upon graduation, he became a Sloan Postdoctoral Fellow in the section of ecology and systematics at Cornell University in New York, then started his faculty career as an assistant professor at the University of Colorado before joining Harvard University in 1995 as assistant professor of organismal and evolutionary biology and assistant curator of the Museum of Comparative Zoology. In 1998, he became John L. Loeb associate professor in the natural sciences, organismic and evolutionary biology at Harvard and accepted a promotion to associate curator.

Farrell has received a number of honors, awards, and grants during his career, including a Swiss National Fellowship in 1994–95, and expedition funding for fieldwork in the Caribbean, Argentina, and North America. He was an associate editor of *Systematic Biology*, a panel member for National Science Foundation programs, and has organized or co-organized many symposia and workshops.

References
Farrell, B., biographical sketch provided 29 June 1999.
———, "'Inordinate Fondness' Explained: Why Are There So Many Beetles?" *Science*, 24 July 1998: 555–9.

Catherine Drew Harvell (1954–)

An associate professor of biological sciences and curator of invertebrates at Cornell University in New York, C. Drew Harvell is concerned about the health of the Florida Keys' corals and the health of the Earth's ecosystems overall.

Harvell began studying sea fan corals in the Keys to investigate their natural disease resistance, such as their antibacterial and antifungal properties, and instead found that the corals in many locations were dying from fungal infections that they previously had survived. Something had caused the corals to become more susceptible to formerly nonfatal infections.

Postdoctoral research associate Kiho Kim presented his and Harvell's findings at the January 1999 meeting of the American Association for the Advancement of Science (AAAS). They and other researchers, including scientists from the University of South Carolina and institutions in Curaçao, the West Indies, Colombia, Puerto Rico, and the British Virgin Islands, also described the extent of the disease in sea fans of the Caribbean in a separate paper in the *Marine Ecology Progress Series*.

Harvell believes that the corals are dying, because they are fighting too many assaults at the same time: fluctuating temperatures, increased pollution, freshwater infiltration, silt runoff, and disease from the common, soil-dwelling *Aspergillus* fungus. These stressors are likely affecting other organisms of the reefs in similar ways, and may herald problems for the Earth's overall ecosystem. In a Cornell press release, Harvell stated:

"Now, one of our jobs is to discover what has compromised the resistance of the corals at some sites. Although a significant number of sea fans have died at a few sites, at many locales they recover from infections, pointing to the success of their natural resistance."

A native of Boston, Harvell earned bachelor's and master's degrees in zoology from the University of Alberta, Edmonton, and a doctorate in zoology from the University of Washington. She was a teaching assistant and co-instructor in field marine ecology at Friday Harbor Laboratories until she took a position in 1986 as assistant professor and curator of invertebrates in the section of ecology and systematics within the division of biological sciences at Cornell. She became an associate professor in 1992.

Associate editor for *American Naturalist* from 1994 to 1998, Harvell has also been a reviewer for various publications, including *Ecology, Evolution, Marine Ecology Progress Series, Science,* and *Limnology and Oceanography.* She received a Young Investigator Award from the Society of American Naturalists in 1986, a Young Faculty Award from the DuPont Foundation in 1986, and a postdoctoral fellowship from the North Atlantic Treaty Organization in 1985. She has organized more than a dozen symposia, including the 1999 AAAS session described above at which Kim presented their results.

She is a member of numerous professional societies, including the American Society of Limnology and Oceanography, Ecological Society of America, International Bryozoological Association, Society for the Study of Evolution, Society of American Naturalists, and Western Society of Naturalists.

References

Cornell University, "Coral Bleaching and Death Could Be Early Warning of Environmental Change, Cornell Ecologists Warn," press release, 22 January 1999.

Harvell, C., curriculum vitae, September 1998.

Nagelkerken, I., et al., "Widespread Disease in Caribbean Sea Fans: II. Patterns of Infection and Tissue Loss," *Marine Ecology Progress Series*, 15 December 1997: 255–263.

Smith, G., et al., "Response of Sea Fans to Infection with *Aspergillus* sp. (Fungi)," *Revista de Biologia Tropical*, 46 Supl., 1998: 205–8.

Thomas C. Kane (1945–)

Five million years after a layer of rock sealed over an area of Romania and left the animals below to "make do" in a world of utter darkness, scientists entered what is now called Movile Cave and saw a very unusual ecosystem. The research team included Serban Sarbu, who went on to study the ecosystem as a doctoral student under biologist Thomas C. Kane at the University of Cincinnati in Ohio. (Sarbu received his Ph.D. in 1996 and has since joined the faculty of a private school in New York.)

The ecosystem provided a view of life forms—insects, spiders, and aquatic animals—that derive their energy not from photosynthesis and the Sun, but from chemosynthesis and hydrogen sulfide.

The cave made news around the world when the researchers asserted that the bases for the ecosystem's food chain were the chemoautotrophic organisms that were able to obtain energy from the hydrogen sulfide and methane that flow into the cave with groundwater. Until the cave's discovery, the only other examples of chemoautotrophic ecosystems were those at the deep-sea vents found at the bottom of the ocean.

The cave's chemoautotrophic organisms serve as the foundation food for the rest of the ecosystem's organisms. The researchers identified four dozen species living in the cave and described nearly three dozen of them as new species. The cave organisms included water scorpions, predatory leeches, and a species of spider in which the individuals lose all eight of their eyes as they mature.

In a press release, Kane noted, "This is a system which is really interesting from a scientific point of view. It's biologically unique. But it serves as a model for bigger issues in biology and has attracted a lot interest."

Kane, who in the summer of 1995 led the first major team of biologists and geologists to explore the Movile Cave, is interested in not only the current species within the ecosystem, but the evolution of troglomorphy, or the morphological condition common to cave-dwelling animals. These organisms typically exhibit degeneration or loss of their eyes, have large antennae and other augmented sensory structures, and have little body color. His work includes biochemical and molecular investigations to study the role of natural selection in the organisms' evolution.

Kane earned a bachelor of science degree in biology from Niagara University in New York, and a master of science and doctorate in biology from the University of Notre Dame in Indiana. He then accepted an assistant professorship at the University of Cincinnati, where he now holds the title of professor of biological sciences. In addition, Kane has served as scientist-in-residence at Adirondack Laboratory in New York, visiting associate professor in the department of ecology and evolutionary biology at Northwestern University in Illinois, visiting associate director of research at Laboratoire Souterrain du CNRS in France, and visiting scholar at the Museum of Zoology at the University of Michigan.

Kane also acts as a consultant for environmental impact questions and a speaker for employee training sessions at Mammoth Cave National Park in Kentucky. A referee for numerous journals, including *Ecology, American Midland Naturalist, Evolution,* and *Journal of Biogeography,* Kane is also a member of various scientific organizations, such as the National Speleological Society, the Society for the Study of Evolution, and the Ecological Society of America.

References

Kane, T. C., biographical sketch, 25 June 1999.

———, curriculum vitae, May 1998.

National Science Foundation, "Buried in Romania: Forever-Dark Cave Crawling with Life," press release, October 1995.

Sarbu, S., and T. Kane, "A Subterranean Chemoautotrophically Based Ecosystem," *The NSS Bulletin*, December 1995: 91–98.

Sarbu, S., et al., "A Chemoautotrophically Based Cave Ecosystem," *Science*, 28 June 1996: 1963–65.

University of Cincinnati, "Surviving Underground: University of Cincinnati Scientists Explore Unusual Romanian Cave System," press release, 13 February 1996.

Eric S. Lander (1957–)

Mathematician-turned-biologist Eric S. Lander is one of the leaders in the drive to sequence the human genome. Lander's scientific group, based at the Whitehead Institute for Biomedical Research in Massachusetts, is one of three selected by the National Human Genome Research Institute to undertake the monumental task of identifying the string of bases within the human genetic code. The other scientists are Robert Waterston and his research group at Washington University in Missouri, and Richard Gibbs' group at Baylor College of Medicine in Texas.

These three teams, along with a research group at the Sanger Centre in England and the Joint Genome Institute of the U.S. Department of Energy, prepared plans to complete a rough draft of the human genome by early 2000 and then finalize it within three years.

Speed become an issue in this research when a private venture announced its intention to sequence the genetic code by 2001. Academic researchers and government officials feared that success by a private company might lead to patents, which could then limit or even halt future genetic research projects (see Chapter 6).

In addition to conducting research at the Whitehead Institute, Lander is the institute's director, a biology professor at the Massachusetts Institute of Technology (MIT), and a geneticist at Massachusetts General Hospital. Despite his current immersion in biological sciences, his education is in mathematics. He earned a A.B. degree with highest honors in mathematics from Princeton University in New Jersey in 1978, and a Ph.D. in mathematics from Oxford University in England in 1981.

His degree from Oxford in hand, Lander took a position as an assistant professor at Harvard University's Graduate School of Business, teaching mathematics, statistics, and economics courses. As he continued that post, he became involved in molecular biology and genetics laboratories at Harvard and MIT, taking positions as a fellow at the Whitehead Institute and a visiting scientist in MIT's biology department.

He eventually left Harvard to join the MIT biology department as an associate professor in 1989. In 1990, he became director of the Whitehead/MIT Center for Genome Research. Three years later, he accepted a promotion to professor at MIT and took his additional appointment at Massachusetts General Hospital.

His litany of honors includes election to the American Academy of Arts and Sciences and the American Academy of Achievement, the American Academy of Microbiology, and the U.S. National Academy of Sciences. He has accepted such awards as the 1998 Pasarow Prize in Cancer, the 1998 Chiron Prize for Biotechnology from the American Society for Microbiology, the 1996 Dickson Prize in Medicine, and the 1996 Rhoads Memorial Award for excellence in cancer research.

He also serves on numerous editorial boards, and as a member of a number of committees, boards, and task forces. They include the scientific advisory boards of Massachusetts General Hospital and Dana-Farber Cancer Institute, the U.S. Presidential Commission on the National Medal of Science, and the Advisory Committee to the Director of the National Institutes of Health.

References

Lander, E. S., curriculum vitae, provided 27 July 1999.

Marshall, E., "NIH to Produce a 'Working Draft' of the Genome by 2001," *Science,* 18 September 1998: 1774–5.

Mullikin, J., and A. McMurray, "Sequencing the Genome, Fast," *Science,* 19 March 1999: 1867–8.

Pennisi, E., "Academic Sequencers Challenge Celera in a Spring to the Finish," *Science,* 19 March 1999: 1822–3.

William H. R. Langridge (1938–)

William H. R. Langridge is leading the way toward genetically engineered food plants that carry vaccines for human diseases. In 1998, he and other scientists in the Center for Molecular Biology and Gene Therapy at Loma Linda University in California, engineered potatoes that fight cholera.

Langridge earned a bachelor's degree in liberal arts and a master's degree in plant genetics from the University of Illinois, and a Ph.D. in biochemistry from the University of Massachusetts. He conducted postdoctoral work in virology and served as assistant scientist in plant molecular biology at the Boyce Thompson Institute at Cornell University in New York, before becoming associate professor in the department of plant science at the University of Alberta in Canada. He left Alberta in 1993 to join Loma Linda's biochemistry department in its Center for Molecular Biology and Gene Therapy.

During his time at Loma Linda, Langridge's research group has built upon his expertise and focused on genetically engineering plants that are more nutritious and that relay disease resistance. Using gene-transfer methods that the group developed, Langridge reported in 1998 that his research team had created genetically engineered potatoes that produced the level of cholera proteins necessary to stimulate the immune system of a mouse to fight the cholera toxin. In other words, the potatoes carried the cholera vaccine, and it worked in mice. His work paved the way for the scientific community to develop inexpensive, edible vaccines for humans.

In creating the vaccine-yielding spuds, Langridge's group inserted a non-pathogenic piece of the cholera B toxin gene into plasmids, or circular pieces of bacterial DNA, and introduced those to the plants. The transgenic potatoes began making the non-pathogenic toxin in higher levels than had previously been seen in other plant vaccine work. When the researchers fed the potatoes to mice, the rodents began making antibodies to the disease and experienced a 60 percent reduction in cholera symptoms. Additional feedings of potatoes served as "booster shots" and made the mice even more cholera-resistant.

While cooking the potatoes reduced the vaccine's potency, the researchers reported that it was still effective in fighting cholera. In addition, their work demonstrated that potatoes could be encouraged to produce high levels of vaccine proteins. They were able to obtain a 0.3 percent level of total soluble protein in the potatoes, they reported—a great improvement over the previous 0.01 percent average in plant-produced proteins.

The Langridge lab's current endeavors include transferring human milk proteins into tomatoes and potatoes. The goal is for the potatoes to begin expressing the proteins. The researchers are also continuing their work toward creating food plants that yield preventive vaccines and that suppress autoimmune diabetes. They also are studying colo-rectal cancer gene therapy.

References

Arakawa, T., et al., "Efficacy of a Food Plant-Based Oral Cholera Toxin B Subunit Vaccine," *Nature Biotechnology*, May 1998: 292–7.

Langridge, W. H. R., biographical sketch, 11 November 1998, <http://www.llu.edu/llu/medicine/biochem/faculty/langridge/htm>, accessed 2 February 2000.

———, curriculum vitae, provided 30 June 1999.

Loma Linda University, "Efficacy of a Food Plant-Based Oral Cholera Toxin B Subunit Vaccine," press release, 26 February 1998, <http://www.llu.edu/llu/medicine/biochem/faculty/langridge.htm>, accessed 2 February 2000.

Sir Robert May (1936–)

A preeminent scientist known around the world for his work in ecosystems and ecological theory, Sir Robert May began a five-year post in

1995 as chief scientific advisor to the United Kingdom government and head of the U.K. Office of Science and Technology. In U.S. terms, the scope of the Office of Science and Technology is similar to that of the National Science Foundation and the nonclinical portion of the National Institutes of Health combined.

Although May started his education in chemical engineering—eventually earning degrees in theoretical physics and lecturing in applied mathematics—biologists count him as one of their own.

May earned bachelor's and doctoral degrees in theoretical physics from Sydney University, then went on to serve as Gordon MacKay Lecturer in Applied Mathematics at Harvard University in Massachusetts and senior lecturer in theoretical physics at Sydney, before taking an interest in animal population dynamics and natural communities. His career shifted further toward biology in 1973 when he accepted a position as Class of 1877 Professor of Zoology at Princeton University in New Jersey. Four years later, he became chair of Princeton's University Research Board, a title he held until he returned to Britain and a Royal Society Research professorship at Oxford University. He is on leave from Oxford and Imperial College while he serves in his current appointments.

Much of May's research revolves around whether complexity in the form of increased species diversity and interaction promotes stability within an ecosystem. Tapping into his wide-ranging expertise, he has compared mathematical models of complex and simple ecosystems and showed that complex systems were likely to be "more dynamically fragile." This finding was in opposition to the commonly held hypothesis that complexity begets stability. In *Ecology and Theories and Applications*, he remarked, "This work, I believe, has refocused the subsequent agenda in this area, as people have become more careful about distinguishing the productivity of a community as a whole from the fluctuations in individual populations and more broadly have sought to understand the intricate patterns that particular ecosystems have woven, over evolutionary time, in ways that undercut any glib generalizations."

He has also been one of several scientists who helped to meld chaos theory and associated mathematical models with studies of biological populations. His interests then led him to the interplay between infectious diseases, and the population levels and distribution of organisms, and he is now studying conservation biology and theoretical immunology.

His background and his bent toward the biological sciences led him to author or to edit numerous books, including *Stability and Complexity in Model Ecosystems, Infectious Diseases of Humans: Transmission and Control, Perspectives in Ecological Theory,* and *Evolution of Biological Diversity.* He has received numerous honors, including election to the

Royal Society in 1979 and to the U.S. National Academy of Sciences as a foreign member in 1992, and a knighthood in 1995. Sir Robert has also received awards from such organizations at the Linnean Society, American Ecological Society, and the Royal Swedish Academy of Science. The latter presented May with its prestigious Crafoord Prize in 1996, citing the scientist's "pioneering ecological research in theoretical analysis of the dynamics of populations, communities and ecosystems."

References

May, R., biographical sketch provided 16 April 1999.
Stiling, P., ed., *Ecology and Theories and Applications*, Upper Saddle River, NJ: Prentice Hall, 1998: 70–71.

Baldomero Olivera (1941–)

In 1996, the work of biologist Baldomero "Toto" Olivera and his colleagues made the cover of the widely respected journal *Nature*. In the cover story, Olivera described how a tropical marine snail uses its poisons to bring fish to a dead stop in as little as one second. The poisons are powerful enough to kill a human. The cone snail, *Conus* spp., delivers the mix of poisons, some of which are the most potent in all of nature, with a harpoon from its long proboscis.

While the natural history of the snails is impressive in itself, Olivera is more interested in the possible medical benefits associated with their poisons. He was able to isolate one of the toxins, omega-conotoxin, from the cone snail *Conus magus*. In the snail's prey, the toxin arrests the transmission of nerve signals by blocking calcium channels in the membrane of the nerve cells.

Olivera realized that the toxin might be of use to treat stroke in humans. Following a stroke, a portion of the brain no longer receives blood or oxygen, and the cells die because they don't have the energy, imparted by the oxygen, to close their calcium channels. Olivera hypothesized that the snail's toxin might be able to temporarily shut down the calcium channels, preventing the essentially irreversible cell death. Other researchers tested the idea and found that the toxin was capable of slowing or halting the death of nerve cells, or neurons, in rats.

In addition, hundreds of research laboratories began studying the venom's components, also called conopeptides, for their ability to selectively target only the intended cell membrane receptors. Olivera and other scientists believe the conopeptides might be useful in developing drugs that have very specific responses, and thus fewer side effects.

Most recently, a small study by the Pain Clinic at Stanford University in California showed that a synthetic drug derived from a cone snail toxin acted as a pain killer by interrupting pain signals. The drug eliminated the long-term pain in several patients, but didn't carry the side effects of the opiate-based painkillers commonly prescribed to patients enduring long-term pain.

Olivera earned a bachelor of science degree in chemistry from the University of the Philippines and a Ph.D. in chemistry from the California Institute of Technology (Caltech) before beginning a postdoctoral appointment at Stanford University. Upon completing the appointment, he returned to the Philippines as a research associate professor of biochemistry at the University of Philippines Medical School. In 1970, he joined the University of Utah as an associate professor of biology and accepted a promotion to professor in 1973. In 1992, he became a distinguished professor of biology at the university and also served from 1994 to 1995 as director of the university's Interdepartmental Neuroscience Program.

In just the last 10 years, Olivera received the Utah Governor's Medal for Science and Technology, and has been Stetten Lecturer at the National Institute of General Medical Sciences, E. E. Just Lecturer at the American Society of Cell Biology, and Cooper Lecturer at Yale University in Connecticut.

For the National Institutes of Health, he was a member of the National Institutes of Health Biochemistry Study Section, of the Advisory Committee to the Director, and of the Cellular and Molecular Basis of Disease Review Committee. He has also served on the editorial boards of *Journal of Biological Chemistry, Journal of Toxicology—Toxin Reviews,* and *Molecular Diversity*, and has been a member of the National Advisory General Medical Sciences Council. He is currently a member of the Burroughs Wellcome Foundation's Toxicology Advisory Board.

References
"Killer Snails, Healer Snails," *Discover*, May 1994: 19–20.

Olivera, B., curriculum vitae, provided in June 1999.

Thalman, J., "Sea Snail's Chemical Weapon Examined As Human Pain Reliever," press release, University of Utah, 2 June 1999.

University of Utah, "U. Biologist: Cone Snail Venom Could Lead to Safer Drugs," press release, 21 January 1994.

Grace Panganiban (1962–)

By studying gene expression and the development of appendages in various animals, Grace Panganiban, Steven Irvine, and a team of researchers in 1997 provided evidence that many of the diverse animals that exist today shared a common ancestor some 600 million years ago.

In particular, the research team studied the expression of the Dll gene in arthropods and the related Dlx genes in other organisms. These genes serve a regulatory role in limb formation in protostomes, such as arthropods, annelids, and molluscs, and in deuterostomes such as mammals, birds, and reptiles.

Using an antibody that stains cells in which the genes were expressed, the researchers surveyed the Dll and related genes in a wide range of animals, including arthropods; fish; polychaetes, or bristleworms; primitive, slug-like onychophores, which are found in the tropics; tunicates; and echinoderms, like sea urchins. In each case the Dll and related genes became activated during limb development.

These results indicated that even though the appendages of insects, birds, humans, and many other animals are widely disparate, limbs did not evolve independently many times. Instead, a single ancestor in Precambrian evolutionary history gave rise to the organisms, and the array of limbs branched out from that single ancestor.

Besides this research, Panganiban, an assistant professor in the department of anatomy at the University of Wisconsin Medical School, is more broadly interested in cell biology and the genetics of animal development. Her work with limb development and evolution began as a postdoctoral fellow in the laboratory of Sean Carroll, also at the University of Wisconsin. During her postdoctoral appointment, Panganiban also studied portions of the fruit fly (*Drosophila*) central nervous system. Following that appointment, she accepted the position of assistant professor at the university.

The anatomist earned her bachelor of arts degree in biology from Swarthmore College in Pennsylvania and her Ph.D. in cellular and molecular biology from the University of Wisconsin–Madison. Her graduate research included a biochemical analysis of the fruit fly protein known as dpp, or decapentaplegic, and an investigation into its developmental role.

During her career as student, postdoctoral fellow, and faculty member, Panganiban has received several honors and awards. She was a National Institutes of Health predoctoral trainee, and she received fellowships from both the NIH and the National Science Foundation/Sloan Foundation. She also accepted an Inbusch Award for Medical Research and a Sloan Foundation Young Investigator appointment, and participated in a National Academy of Sciences invited symposium.

In addition, Panganiban continues to spend time as a tutor and mentor for inner-city middle and high school students, has hosted an enrichment program for middle school students, and has assisted with a genetics program for high school students.

References

Panganiban, G., curriculum vitae, provided 6 July 1999.

Panganiban, G., et al., "The Origin and Evolution of Animal Appendages," *Proceedings of the National Academy of Sciences*, May 1997: 5162–6.

Zimmer, C., "Hidden Unity," *Discover*, January 1998: 46–47.

Steward T. A. Pickett (1950–)

Steward T. A. Pickett is a botanist on a mission. With backgrounds in biodiversity, the ecology of gradients between urban and rural environments, natural disturbance, and the dynamics of plant community, Pickett is now director of the Baltimore Ecosystem Study. The study is one of the National Science Foundation's 20 Long-Term Ecological Research (LTER) sites. Through the Baltimore study, titled "Human Settlements as Ecosystems: Metropolitan Baltimore from 1797–2100," scientists hope to learn how changes to an urban environment can alter the socioeconomic and ecological outcomes of the area. The Baltimore study also has a strong educational focus in cooperation with area schools, and a watershed component, which is funded by the Environmental Protection Agency. Pickett noted, "This project builds on a remarkable foundation of social science research, a long history of applying community forestry to issues of ecological health of the Baltimore metropolitan area and the Chesapeake Bay, and excellent relationships between the community and research in the area."

Kentucky born, Pickett earned a bachelor of science degree in botany from the University of Kentucky in 1972, followed by a Ph.D. in botany in 1977 from the University of Illinois at Urbana-Champaign. For his doctorate, he specialized in plant ecology. Upon graduation, he accepted a position on the faculty of Rutgers University in New Jersey, and a few years later also became director of Rutgers' Hutcheson Memorial Forest Center. Pickett left Rutgers in 1987 for the Institute of Ecosystem Studies, where he had been serving as visiting scientist. He is currently senior scientist at the institute, in addition to director of the Baltimore Ecosystem Study and adjunct professor at the University of Connecticut.

In his research, Pickett has conducted experiments on the roles of forest edges and patchiness in ecosystems and, in particular, on species diversity and productivity. His work on the Baltimore study is especially reflected in his research into cities as ecological systems, and into vegetation dynamics and natural disturbance.

A fellow of both the American Association for the Advancement of Science and the American Academy of Arts and Sciences, Pickett has also held numerous professional titles. These include vice president for science of the Ecological Society of America, member of the National Research Council's Committee on Scientific Issues in the Endangered Species Act for the National Research Council, member of both the board of directors and the scientific advisory board of the Defenders of Wildlife, and member of the board of directors of the Center for Ecological Synthesis and Analysis of the University of California–Santa Barbara.

In addition to more than 100 scientific papers, Pickett has coedited four books. His *The Ecology of Patch Dynamics and Natural Disturbance* is often cited as a classic in the field. His other books include *Ecological Heterogeneity, Humans as Components of Ecosystems,* and *The Ecological Basis of Conservation.*

References

Baltimore Ecosystem, Long-Term Ecological Research, descriptive handout from "The Metropolis in the Millennium," a session of the annual meeting of the American Association for the Advancement of Science, Anaheim, CA, 23 January 1999.

Pickett, S. T. A.; biographical sketch provided at "The Metropolis in the Millennium," a session of the annual meeting of the American Association for the Advancement of Science, Anaheim, CA, 23 January 1999.

Eric Sanford (1968–)

Eric Sanford was still working toward his doctoral degree in marine ecology when one of his papers landed in the major journal *Science.* The paper explained how small climate changes could trigger large modifications within natural communities. In particular, he studied the impact of shifts in water temperature on the relationship between the sea star *Pisaster ochraceus* and the rocky intertidal mussels *Mytilus californianus* and *M. trossulus.* The sea star is a keystone predator, and its primary prey are the two mussels.

His findings indicated that temperature changes of as little as 3°C could cause the sea stars to stop feeding on the mussels, which then had the opportunity to overpopulate and dominate the ecosystem, perhaps leading to the extirpation of other species. The work was at odds with previously held beliefs indicating that temperature changes would have more gradual effects on ecosystems.

Sanford, at the time a Ph.D. candidate with an expected graduation date of September 1999, was working under coadvisers Bruce Menge and Jane Lubchenco in Oregon State University's zoology department. (Lubchenco is a well-known biologist and past president of the American Association for the Advancement of Science.) Sanford finished his bachelor of arts degree in biology at Brown University in Rhode Island, but completed two semesters of his undergraduate education at the University of California–Santa Cruz studying marine biology and behavioral ecology.

His research experience includes stints as a laboratory assistant at the University of California–Santa Cruz studying the effects of photoperiod on reproductive ecology, a research assistant at Brown University investigating how water temperature and flow affect acorn barnacles, a scientific aide at the California Department of Fish and Game studying the relationships between freshwater outflow on fish and shrimp, and a research assistant at Oregon State University analyzing benthic intertidal communities in New Zealand.

He has garnered a number of grants and fellowships throughout his educational career. He accepted a National Science Foundation Predoctoral Fellowship for 1993–97 at Oregon State University, a 1997 Lerner-Gray Fund for Marine Research grant from the American Museum of Natural History, and a National Wildlife Federation Climate Change Fellowship in 1998. His publications include articles in the *Marine Ecology Progress Series, Proceedings of the National Academy of Sciences, Science, Ecological Monographs* and *Biological Bulletin.* He has also presented his work at symposia, conferences, and meetings in Oregon, New Mexico, California, Florida, Washington, and Baja Mexico.

References

Sanford, E., curriculum vitae.

——, "Regulation of Keystone Predation by Small Changes in Ocean Temperature," *Science*, 26 March 1999: 2095–7.

Stanley K. Sessions (1949–)

Stanley Sessions became interested in the causes of amphibian deformities, such as additional legs, when colleague Steve Ruth came across a northern California pond filled with thousands of deformed Pacific treefrogs (*Hyla regilla*) and long-toed salamanders (*Ambystoma macrodactylum*). Because the species fell in two separate orders of the amphibians, they looked to the environment rather than genetic abnormalities for the cause.

After testing the water for chemical pollution and detecting nothing that would result in the deformities, they stained the malformed specimens and found that each carried an infection of parasitic flatworms called trematodes, and each had cysts around and in the hind limbs or in the hind limb buds (the developing limbs). The cysts appeared to be closely associated with such deformities as missing limbs or limb structures, fused skin, or multiple limbs.

Sessions reported that the amphibians are just one part of the trematode's life cycle. Aquatic snails pick up trematode eggs which hatch into larvae and then exit the snails as swimming larvae. At this stage, the trematodes enter the amphibians to form the cysts. The flatworm's life cycle is completed when garter snakes (*Thamnophis* spp.) eat the amphibians and become the hosts for the adult flatworms.

Sessions first reported the trematode-amphibian relationship in 1990 and described additional findings at the American Association for the Advancement of Science meeting in January 1999. His work received nationwide media coverage in the spring of 1999 when he published another paper about the amphibian-trematode connection in the journal *Science*.

Sessions's research group is continuing its work to determine causes for the full range of amphibian deformities, is trying to develop a way to predict future epidemics of multilegged frogs, and is beginning to investigate the potential co-evolution of the trematodes and amphibians.

Sessions earned a bachelor of arts degree in anthropology from the University of Oregon and a doctorate in zoology from the University of California–Berkeley, where he was a research and teaching assistant, and a teaching associate. He also was a research associate for the department of zoology at the University of Leicester in England. He joined Hartwick College's biology department in 1989 as an assistant professor and accepted the title of associate professor there in 1996.

A member of several organizations, including the American Society of Ichthyologists and Herpetologists, Herpetologists' League, Society for Developmental Biology, and Society for the Study of Evolution, Sessions has also served on the editorial board of *Chromosome Research* and as corresponding editor of the *Journal of Evolutionary Biology*.

References

Sessions, S., curriculum vitae, <http://www.hartwick.edu/biology/def_frogs/stan/stan.html>, accessed 2 February 2000.

———, "'Yuck, Gross!': What Can Online Deformed Frogs Teach About Science?" session at the American Association for the Advancement of Science 1999 annual meeting, Anaheim, CA, 25 January 1999.

Ian Wilmut (1944–)

The world's eyes turned toward Ian Wilmut after he led a scientific team in cloning the first animal from adult cells, a feat that many scientists believed was impossible.

Wilmut and his team at the Roslin Institute in Scotland announced in 1997 to a surprised and, at first, skeptical scientific community that they had been successful in their efforts to create a lamb from the cell of an adult ewe's mammary gland. The lamb was born in July 1996. A key to generating the clone was a technique known as nuclear transfer, or the removal of a nucleus from one organism's cell and its replacement with the nucleus from the cell of another organism.

Another important step in the creation of the cloned lamb, Dolly, was encouraging the six-year-old adult cell's nucleus to "act" like the nucleus in egg cells. Egg cells are totipotent—they have the ability to differentiate into nearly any other type of cell. Once a cell is differentiated, however, its fate is normally sealed, and it cannot become another type of cell. Differentiation, then, is normally irreversible. The Roslin researchers reasoned that starvation might provoke the adult nuclei to enter an inactive state that would revive its totipotency, so they starved the mammary cells for five days. The technique apparently worked and the nuclei became totipotent, but the scientists still weren't sure exactly what mechanisms were involved in the nuclear reprogramming.

(As mentioned elsewhere in this book, Dolly isn't a true clone, because she isn't genetically identical to the "mother." While Dolly received the vast majority of the mother's genetic component from her cell nucleus, Dolly's mitochondrial genes came from the recipient cell, not the donor nucleus.)

Although Wilmut explained to the media the importance of cloning in generating genetically identical animals that carry genes useful in treating human disease, the announcement about Dolly quickly shifted to debates over whether humans could be cloned and how that eventuality could be prevented. Since the announcement, other reports of cloned animals have appeared in the scientific journals, and a great deal of research is under way to understand the mechanisms involved in the creation of Dolly.

Wilmut received his B.Sc. degree in agricultural science from the University of Nottingham and his doctorate from the University of Cambridge, both of which are in England. His doctoral research was titled "Deep Freeze Preservation of Boar Semen." He continued his studies and eventually was able to entice a calf embryo from the freezer to a successful birth. The resulting calf, named "Frosty," represented the first time such work had been successful.

Wilmut conducted research at the Animal Breeding Research Organization in Edinburgh before joining the Roslin Institute as a principal investigator. Wilmut is also the chief scientific officer of Geron Bio-Med. In his current positions, he is continuing to study the nuclear transfer procedure and associated biomedical applications.

Since the birth of Dolly, Wilmut has received a number of honorary doctorates, as well as a Golden Plate Award in 1998, the Lord Lloyd Kilgerran Prize in 1998, the Sir John Hammond Memorial Prize in 1998, and the 1999 Sir William Young Award.

References

Pennisi, E., "The Lamb That Roared," *Science,* 19 December 1997, 2038–9.
Travis, J., "Ewe Again? Cloning from Adult DNA," *Science News,* 1 March 1997, 132.
———, "A Fantastical Experiment," *Science News,* 5 April 1997, 214.
Wilmut, I.; biographical sketch, provided 14 July 1999
Wills, C., "A Sheep in Sheep's Clothing?" *Discover,* January 1998, 22–23.

CHAPTER EIGHT
Documents, Letters, and Reports

This chapter provides a glimpse of the exchanges various biology topics have spawned. It begins with documents that relate to the environment, including selected portions of the U.S. Endangered Species Act. Because of the furor over biological research that has direct ties to humans, the remainder of this chapter will focus on governmental-related discussions on those topics, including the protection of human research subjects, genetic testing, cloning, and stem cell research.

ENVIRONMENTAL TOPICS

Environmental topics have generated lively debates. One topic that continues to be discussed is the Endangered Species Act of 1973, which provides legal protection to "fish, wildlife, and plant" species deemed to be threatened with extinction. Current information about policies and new listing decisions are available at <http://endangered.fws.gov/>. Another important environmental document is the Kyoto Protocol, which requires, in part, that industrialized nations reduce greenhouse gases to a level at least 5 percent below that of 1990. The text of the Kyoto Protocol to the United Nations (1997) is not included here, but is available in its entirety at <http://www.cnn.com/SPECIALS/1997/global.warming/stories/treaty/>.

Endangered Species Act

The Endangered Species Act (ESA) was designed to protect species threatened with extinction. It requires lawmakers, property developers, biologists and other scientists, business people, and all other citizens to take into account those species and their associated ecosystems before taking actions that might harm them. Although the act was roundly approved in 1973—unanimously in the Senate and nearly unanimously in the House of Representatives—it has generated a great deal of controversy since. Few imagined that the number of species under its protection would grow from 109 in 1973 to more than 10 times that number less than three decades later, or that the majority of the endangered species on its list would be plants. From 1973 to 1999, 11 species had recovered to such a degree that they were removed from the list.

Advocates hail the successes and note that the act has brought attention to all declining species, including those not on the endangered species list, while protecting important ecosystems. Detractors focus on the number of failures. They claim the law is putting unnecessary restrictions on the rights of property owners whose land is deemed "critical habitat" for a species listed as endangered, and that it is also taking away jobs by preventing such activities as logging in protected areas.

Endangered Species Act of 1973 (selected sections)

Section 2. Congressional findings and declaration of purposes and policy

(a) Findings

   The Congress finds and declares that

(1) various species of fish, wildlife, and plants in the United States have been rendered extinct as a consequence of economic growth and development untempered by adequate concern and conservation;

(2) other species of fish, wildlife, and plants have been so depleted in numbers that they are in danger of or threatened with extinction;

(3) these species of fish, wildlife, and plants are of aesthetic, ecological, educational, historical, recreational, and scientific value to the Nation and its people;

(4) the United States has pledged itself as a sovereign state in the international community to conserve to the extent practicable the various species of fish or wildlife and plants facing extinction, pursuant to:

(A) migratory bird treaties with Canada and Mexico;

(B) the Migratory and Endangered Bird Treaty with Japan;

(C) the Convention on Nature Protection and Wildlife Preservation in the Western Hemisphere;

(D) the International Convention for the Northwest Atlantic Fisheries;

(E) the International Convention for the High Seas Fisheries of the North Pacific Ocean;

(F) the Convention on International Trade in Endangered Species of Wild Fauna and Flora; and

(G) other international agreements; and

(5) encouraging the States and other interested parties, through Federal financial assistance and a system of incentives, to develop and maintain conservation programs which meet national and international standards is a key to meeting the Nation's international commitments and to better safeguarding, for the benefit of all citizens, the Nation's heritage in fish, wildlife, and plants.

(b) Purposes

The purposes of this chapter are to provide a means whereby the ecosystems upon which endangered species and threatened species depend may be conserved, to provide a program for the conservation of such endangered species and threatened species, and to take such steps as may be appropriate to achieve the purposes of the treaties and conventions set forth in subsection (a) of this section.

(c) Policy

(1) It is further declared to be the policy of Congress that all Federal departments and agencies shall seek to conserve endangered species and threatened species and shall utilize their authorities in furtherance of the purposes of this chapter.

(2) It is further declared to be the policy of Congress that Federal agencies shall cooperate with State and local agencies to resolve water resource issues in concert with conservation of endangered species.

. . .

Section 4. Determination of endangered species and threatened species

(a) General

(1) The Secretary shall by regulation promulgated in accordance with subsection (b) of this section determine whether any species is an endangered species or a threatened species because of any of the following factors:

(A) the present or threatened destruction, modification, or curtailment of its habitat or range;

(B) overutilization for commercial, recreational, scientific, or educational purposes;

(C) disease or predation;

(D) the inadequacy of existing regulatory mechanisms; or

(E) other natural or manmade factors affecting its continued existence.

(2) With respect to any species over which program responsibilities have been vested in the Secretary of Commerce pursuant to Reorganization Plan Numbered 4 of 1970

(A) in any case in which the Secretary of Commerce determines that such species should

(i) be listed as an endangered species or a threatened species, or

(ii) be changed in status from a threatened species to an endangered species, he shall so inform the Secretary of the Interior, who shall list such species in accordance with this section;

(B) in any case in which the Secretary of Commerce determines that such species should

(i) be removed from any list published pursuant to subsection (c) of this section, or

(ii) be changed in status from an endangered species to a threatened species, he shall recommend such action to the Secretary of the Interior, and the Secretary of the Interior, if he concurs in the recommendation, shall implement such action; and

(C) the Secretary of the Interior may not list or remove from any list any such species, and may not change the status of any such species which are listed, without a prior favorable determination made pursuant to this section by the Secretary of Commerce.

(3) The Secretary, by regulation promulgated in accordance with subsection (b) of this section and to the maximum extent prudent and determinable

(A) shall, concurrently with making a determination under paragraph (1) that a species is an endangered species or a threatened species, designate any habitat of such species which is then considered to be critical habitat; and

(B) may, from time-to-time thereafter as appropriate, revise such designation.

(b) Basis for determinations

(1)(A) The Secretary shall make determinations required by subsection (a) (1) of this section solely on the basis of the best scientific and commercial data available to him after conducting a review of the status of the species and after taking into account those efforts, if any, being made by any State or foreign nation, or any political subdivision of a State or foreign nation, to protect such species, whether by predator control, protection of habitat and food supply, or other conservation practices, within any area under its jurisdiction, or on the high seas.

(B) In carrying out this section, the Secretary shall give consideration to species which have been

(i) designated as requiring protection from unrestricted commerce by any foreign nation, or pursuant to any international agreement; or

(ii) identified as in danger of extinction, or likely to become so within the foreseeable future, by any State agency or by any agency of a foreign nation that is responsible for the conservation of fish or wildlife or plants.

(2) The Secretary shall designate critical habitat, and make revisions thereto, under subsection (a)

(3) of this section on the basis of the best scientific data available and after taking into consideration the economic impact, and any other relevant impact, of specifying any particular area as critical habitat. The Secretary may exclude any area from critical habitat if he determines that the benefits of such exclusion outweigh the benefits of specifying such area as part of the critical habitat, unless he determines, based on the best scientific and commercial data available, that the failure to designate such area as critical habitat will result in the extinction of the species concerned.

(3)(A) To the maximum extent practicable, within 90 days after receiving the petition of an interested person under section 553(e) of Title 5 to add a species to, or to remove a species from, either of the lists published under subsection (c) of this section, the Secretary shall make a finding as to whether the petition presents substantial scientific or commercial information indicating that the petitioned action may be warranted. If such a petition is found to present such information, the Secretary shall promptly commence a review of the status of the species concerned. The Secretary shall promptly publish each finding made under this subparagraph in the Federal Register.

(B) Within 12 months after receiving a petition that is found under subparagraph (A) to present substantial information indicating that the petitioned action may be warranted, the Secretary shall make one of the following findings:

(i) The petitioned action is not warranted, in which case the Secretary shall promptly publish such finding in the Federal Register.

(ii) The petitioned action is warranted, in which case the Secretary shall promptly publish in the Federal Register a general notice and the complete text of a proposed regulation to implement such action in accordance with paragraph (5).

(iii) The petitioned action is warranted, but that

(I) the immediate proposal and timely promulgation of a final regulation implementing the petitioned action in accordance with paragraphs (5) and (6) is precluded by pending proposals to determine whether any species is an endangered species or a threatened species, and

(II) expeditious progress is being made to add qualified species to either of the lists published under subsection (c) of this section and to remove from such lists species for which the protections of this chapter are no longer necessary, in which case the Secretary shall promptly publish such finding in the Federal Register, together with a description and evaluation of the reasons and data on which the finding is based.

(C) (i) A petition with respect to which a finding is made under subparagraph (B) (iii) shall be treated as a petition that is resubmitted to the Secretary under subparagraph (A) on the date of such finding and that presents substantial scientific or commercial information that the petitioned action may be warranted.

(ii) Any negative finding described in subparagraph (A) and any finding described in subparagraph (B) (i) or (iii) shall be subject to judicial review.

(iii) The Secretary shall implement a system to monitor effectively the status of all species with respect to which a finding is made under subparagraph (B) (iii) and shall make prompt use of the authority under paragraph 7 to prevent a significant risk to the well being of any such species.

(D) (i) To the maximum extent practicable, within 90 days after receiving the petition of an interested person under section 553(e) of Title 5, to revise a critical habitat designation, the Secretary shall make a finding as to whether the petition presents substantial scientific information indicating that the revision may be warranted. The Secretary shall promptly publish such finding in the Federal Register.

(ii) Within 12 months after receiving a petition that is found under clause (i) to present substantial information indicating that the requested revision may be warranted, the Secretary shall determine how he intends to proceed with the requested revision, and shall promptly publish notice of such intention in the Federal Register.

(4) Except as provided in paragraphs (5) and (6) of this subsection, the provisions of section 553 of Title 5 (relating to rule-making procedures), shall apply to any regulation promulgated to carry out the purposes of this chapter.

(5) With respect to any regulation proposed by the Secretary to implement a determination, designation, or revision referred to in subsection (a) (1) or (3) of this section, the Secretary shall -

(A) not less than 90 days before the effective date of the regulation -

(i) publish a general notice and the complete text of the proposed regulation in the Federal Register, and

(ii) give actual notice of the proposed regulation (including the complete text of the regulation) to the State agency in each State in which the species is believed to occur, and to each county or equivalent jurisdiction in which the species is believed to occur, and invite the comment of such agency, and each such jurisdiction, thereon;

(B) insofar as practical, and in cooperation with the Secretary of State, give notice of the proposed regulation to each foreign nation in which the species is believed to occur or whose citizens harvest the species on the high seas, and invite the comment of such nation thereon;

(C) give notice of the proposed regulation to such professional scientific organizations as he deems appropriate;

(D) publish a summary of the proposed regulation in a newspaper of general circulation in each area of the United States in which the species is believed to occur; and

(E) promptly hold one public hearing on the proposed regulation if any person files a request for such a hearing within 45 days after the date of publication of general notice.

(6)
(A) Within the one-year period beginning on the date on which general notice is published in accordance with paragraph (5) (A) (i) regarding a proposed regulation, the Secretary shall publish in the Federal Register

(i) if a determination as to whether a species is an endangered species or a threatened species, or a revision of critical habitat, is involved, either -

(I) a final regulation to implement such determination,

(II) a final regulation to implement such revision or a finding that such revision should not be made,

(III) notice that such one-year period is being extended under subparagraph (B) (i), or

(IV) notice that the proposed regulation is being withdrawn under subparagraph (B) (ii), together with the finding on which such withdrawal is based; or

(ii) subject to subparagraph (C), if a designation of critical habitat is involved, either

(I) a final regulation to implement such designation, or

(II) notice that such one-year period is being extended under such subparagraph.

(B) (i) If the Secretary finds with respect to a proposed regulation referred to in subparagraph (A) (i) that there is substantial disagreement regarding the sufficiency or accuracy of the available data relevant to the determination or revision concerned, the Secretary may extend the one-year period specified in subparagraph (A) for not more than six months for purposes of soliciting additional data.

(ii) If a proposed regulation referred to in subparagraph (A) (i) is not promulgated as a final regulation within such one-year period (or longer period if extension under clause (i) applies) because the Secretary finds that there is not sufficient evidence to justify the action proposed by the regulation, the Secretary shall immediately withdraw the regulation. The finding on which a withdrawal is based shall be subject to judicial review. The Secretary may not propose a regulation that has previously been withdrawn under this clause unless he determines that sufficient new information is available to warrant such proposal.

(iii) If the one-year period specified in subparagraph (A) is extended under clause (i) with respect to a proposed regulation, then before the close of such extended period the Secretary shall publish in the Federal Register either a final regulation to implement the determination or revision concerned, a finding that the revision should not be made, or a notice of withdrawal of the regulation under clause (ii), together with the finding on which the withdrawal is based.

(C) A final regulation designating critical habitat of an endangered species or a threatened species shall be published concurrently with the final regulation implementing the determination that such species is endangered or threatened, unless the Secretary deems that

(i) it is essential to the conservation of such species that the regulation implementing such determination be promptly published; or

(ii) critical habitat of such species is not then determinable, in which case the Secretary, with respect to the proposed regulation to designate such habitat, may extend the one-year period specified in subparagraph (A) by not more than one additional year, but not later than the close of such additional year the Secretary must publish a final regulation, based on such data as may be available at that time, designating, to the maximum extent prudent, such habitat.

(7) Neither paragraph (4), (5), or (6) of this subsection nor section 553 of Title 5 shall apply to any regulation issued by the Secretary in regard to any emergency posing a significant risk to the well-being of any species of fish or wildlife or plants, but only if -

(A) at the time of publication of the regulation in the Federal Register the Secretary publishes therein detailed reasons why such regulation is necessary; and

(B) in the case such regulation applies to resident species of fish or wildlife, or plants, the Secretary gives actual notice of such regulation to the State agency in each State in which such species is believed to occur.

Such regulation shall, at the discretion of the Secretary, take effect immediately upon the publication of the regulation in the Federal Register. Any regulation promulgated under the authority of this paragraph shall cease to have force and effect at the close of the 240-day period following the date of publication unless, during such 240-day period, the rule-making procedures which would apply to such regulation without regard to this paragraph are complied with. If at any time after issuing an emergency regulation the Secretary determines, on the basis of the best appropriate data available to him, that substantial evidence does not exist to warrant such regulation, he shall withdraw it.

(8) The publication in the Federal Register of any proposed or final regulation which is necessary or appropriate to carry out the purposes of this chapter shall include a summary by the Secretary of the data on which such regulation is based and shall show the relationship of such data to such regulation; and if such regulation designates or revises critical habitat, such summary shall, to the maximum extent practicable, also include a brief description and evaluation of those activities (whether public or private) which, in the opinion of the Secretary, if undertaken may adversely modify such habitat, or may be affected by such designation.

(c) Lists

(1) The Secretary of the Interior shall publish in the Federal Register a list of all species determined by him or the Secretary of Commerce to be endangered species and a list of all species determined by him or the Secretary of Commerce to be threatened species. Each list shall refer to the species contained therein by scientific and common name or names, if any, specify with respect to each such species over what portion of its range it is endangered or threatened, and specify any critical habitat within such range. The Secretary shall from time to time revise

each list published under the authority of this subsection to reflect recent determinations, designations, and revisions made in accordance with subsections (a) and (b) of this section.

(2) The Secretary shall

(A) conduct, at least once every five years, a review of all species included in a list which is published pursuant to paragraph (1) and which is in effect at the time of such review; and

(B) determine on the basis of such review whether any such species should -

(i) be removed from such list;

(ii) be changed in status from an endangered species to a threatened species; or

(iii) be changed in status from a threatened species to an endangered species. Each determination under subparagraph (B) shall be made in accordance with the provisions of subsections (a) and (b) of this section.

(d) Protective regulations

Whenever any species is listed as a threatened species pursuant to subsection (c) of this section, the Secretary shall issue such regulations as he deems necessary and advisable to provide for the conservation of such species. The Secretary may by regulation prohibit with respect to any threatened species any act prohibited under section 1538(a) (1) of this title, in the case of fish or wildlife, or section 1538(a) (2) of this title, in the case of plants, with respect to endangered species; except that with respect to the taking of resident species of fish or wildlife, such regulations shall apply in any State which has entered into a cooperative agreement pursuant to section 1535(c) of this title only to the extent that such regulations have also been adopted by such State.

(e) Similarity of appearance cases

The Secretary may, by regulation of commerce or taking, and to the extent he deems advisable, treat any species as an endangered species or threatened species even though it is not listed pursuant to this section if he finds that -

(A) such species so closely resembles in appearance, at the point in question, a species which has been listed pursuant to such section that enforcement personnel would have substantial difficulty in attempting to differentiate between the listed and unlisted species;

(B) the effect of this substantial difficulty is an additional threat to an endangered or threatened species; and

(C) such treatment of an unlisted species will substantially facilitate the enforcement and further the policy of this chapter.

(f) Recovery plans

(1) The Secretary shall develop and implement plans (hereinafter in this subsection referred to as "recovery plans") for the conservation and survival of endangered species and threatened species listed pursuant to this section, unless he finds that such a plan will not promote the conservation of the species. The Secretary, in developing and implementing recovery plans, shall, to the maximum extent practicable -

(A) give priority to those endangered species or threatened species, without regard to taxonomic classification, that are most likely to benefit from such plans, particularly those species that are, or may be, in conflict with construction or other development projects or other forms of economic activity;

(B) incorporate in each plan -

(i) a description of such site-specific management actions as may be necessary to achieve the plan's goal for the conservation and survival of the species;

(ii) objective, measurable criteria which, when met, would result in a determination, in accordance with the provisions of this section, that the species be removed from the list; and

(iii) estimates of the time required and the cost to carry out those measures needed to achieve the plan's goal and to achieve intermediate steps toward that goal.

(2) The Secretary, in developing and implementing recovery plans, may procure the services of appropriate public and private agencies and institutions, and other qualified persons. Recovery teams appointed pursuant to this subsection shall not be subject to the Federal Advisory Committee Act.

(3) The Secretary shall report every two years to the Committee on Environment and Public Works of the Senate and the Committee on Merchant Marine and Fisheries of the House of Representatives on the status of efforts to develop and implement recovery plans for all species listed pursuant to this section and on the status of all species for which such plans have been developed.

(4) The Secretary shall, prior to final approval of a new or revised recovery plan, provide public notice and an opportunity for public review and comment on such plan. The Secretary shall consider all information presented during the public comment period prior to approval of the plan.

(5) Each Federal agency shall, prior to implementation of a new or revised recovery plan, consider all information presented during the public comment period under paragraph (4).

(g) Monitoring

(1) The Secretary shall implement a system in cooperation with the States to monitor effectively for not less than five years the status of all species which have recovered to the point at which the measures provided pursuant to this chapter are no longer necessary and which, in accordance with the provisions of this section, have been removed from either of the lists published under subsection (c) of this section.

(2) The Secretary shall make prompt use of the authority under paragraph 7 of subsection (b) of this section to prevent a significant risk to the well-being of any such recovered species.

(h) Agency guidelines; publication in Federal Register; scope; proposals and amendments: notice and opportunity for comments

The Secretary shall establish, and publish in the Federal Register, agency guidelines to insure that the purposes of this section are achieved efficiently and effectively. Such guidelines shall include, but are not limited to -

(1) procedures for recording the receipt and the disposition of petitions submitted under subsection (b) (3) of this section;

(2) criteria for making the findings required under such subsection with respect to petitions;

(3) a ranking system to assist in the identification of species that should receive priority review under subsection (a) (1) of this section; and

(4) a system for developing and implementing, on a priority basis, recovery plans under subsection (f) of this section.

The Secretary shall provide to the public notice of, and opportunity to submit written comments on, any guideline (including any amendment thereto) proposed to be established under this subsection.

(i) Submission to State agency of justification for regulations inconsistent with State agency's comments or petition.

If, in the case of any regulation proposed by the Secretary under the authority of this section, a State agency to which notice thereof was given in accordance with subsection (b) (5) (A) (ii) of this section files comments disagreeing with all or part of the proposed regulation, and the Secretary issues a final regulation which is in conflict with such comments, or if the Secretary fails to adopt a regulation pursuant to an action petitioned by a State agency under subsection (b) (3) of this section, the Secretary shall submit to the State agency a written justification for his failure to adopt regulations consistent with the agency's comments or petition.

. . .

Section 6. Cooperation with States

(a) Generally

In carrying out the program authorized by this chapter, the Secretary shall cooperate to the maximum extent practicable with the States. Such cooperation shall include consultation with the States concerned before acquiring any land or water, or interest therein, for the purpose of conserving any endangered species or threatened species.

Section 7. Interagency cooperation

(a) Federal agency actions and consultations

(1) The Secretary shall review other programs administered by him and utilize such programs in furtherance of the purposes of this chapter. All other Federal agencies shall, in consultation with and with the assistance of the Secretary, utilize their authorities in furtherance of the purposes of this chapter by carrying out programs for the conservation of endangered species and threatened species listed pursuant to section 4 of this Act.

(2) Each Federal agency shall, in consultation with and with the assistance of the Secretary, insure that any action authorized, funded, or carried out by such agency (hereinafter in this section referred to as an "agency action") is not likely to jeopardize the continued existence of any endangered species or threatened species or result in the destruction or adverse modification of habitat of such species which is determined by the Secretary, after consultation as appropriate with affected States, to be critical, unless such agency has been granted an exemption for such action by the Committee pursuant to subsection (h) of this section. In fulfilling the requirements of this paragraph each agency shall use the best scientific and commercial data available.

(3) Subject to such guidelines as the Secretary may establish, a Federal agency shall consult with the Secretary on any prospective agency action at the request of, and in cooperation with, the prospective permit or license applicant if the applicant has reason to believe that an endangered species or a threatened species may be present in the area affected by his project and that implementation of such action will likely affect such species.

(4) Each Federal agency shall confer with the Secretary on any agency action which is likely to jeopardize the continued existence of any species proposed to be listed under section 4 of this Act or result in the destruction or adverse modification of critical habitat proposed to be designated for such species. This paragraph does not require a limitation on the commitment of resources as described in subsection (d).

. . .

(c) Biological assessment

(1) To facilitate compliance with the requirements of subsection (a) (2) of this section, each Federal agency shall, with respect to any agency action of such agency for which no contract for construction has been entered into and for which no construction has begun on November 10, 1978, request of the Secretary information whether any species which is listed or proposed to be listed may be present in the area of such proposed action. If the Secretary advises, based on the best scientific and commercial data available, that such species may be present, such agency shall conduct a biological assessment for the purpose of identifying any endangered species or threatened species which is likely to be affected by such action. Such assessment shall be completed within 180 days after the date on which initiated (or within such other period as is mutually agreed to by the

Secretary and such agency, except that if a permit or license applicant is involved, the 180-day period may not be extended unless such agency provides the applicant, before the close of such period, with a written statement setting forth the estimated length of the proposed extension and the reasons therefor) and, before any contract for construction is entered into and before construction is begun with respect to such action. Such assessment may be undertaken as part of a Federal agency's compliance with the requirements of section 102 of the National Environmental Policy Act of 1969 (42 U.S.C. 4332).

(2) Any person who may wish to apply for an exemption under subsection (g) of this section for that action may conduct a biological assessment to identify any endangered species or threatened species which is likely to be affected by such action. Any such biological assessment must, however, be conducted in cooperation with the Secretary and under the supervision of the appropriate Federal agency.

(d) Limitation on commitment of resources

After initiation of consultation required under subsection (a) (2) of this section, the Federal agency and the permit or license applicant shall not make any irreversible or irretrievable commitment of resources with respect to the agency action which has the effect of foreclosing the formulation or implementation of any reasonable and prudent alternative measures which would not violate subsection (a) (2) of this section.

(e) Establishment of committee

(1) There is established a committee to be known as the Endangered Species Committee (hereinafter in this section referred to as the "Committee").

(2) The Committee shall review any application submitted to it pursuant to this section and determine in accordance with subsection (h) of this section whether or not to grant an exemption from the requirements of subsection (a) (2) of this section for the action set forth in such application.

(3) The Committee shall be composed of seven members as follows:

(A) The Secretary of Agriculture.

(B) The Secretary of the Army.

(C) The Chairman of the Council of Economic Advisors.

(D) The Administrator of the Environmental Protection Agency.

(E) The Secretary of the Interior.

(F) The Administrator of the National Oceanic and Atmospheric Administration.

. . .

(h) Exemption

(1) The Committee shall make a final determination whether or not to grant an exemption within 30 days after receiving the report of the Secretary pursuant to

subsection (g) (5) of this section. The Committee shall grant an exemption from the requirements of subsection (a) (2) of this section for an agency action if, by a vote of not less than five of its members voting in person -

(A) it determines on the record, based on the report of the Secretary, the record of the hearing held under subsection (g) (4) of this section and on such other testimony or evidence as it may receive,

that

(i) there are no reasonable and prudent alternatives to the agency action;

(ii) the benefits of such action clearly outweigh the benefits of alternative courses of action consistent with conserving the species or its critical habitat, and such action is in the public interest;

(iii) the action is of regional or national significance; and

(iv) neither the Federal agency concerned nor the exemption applicant made any irreversible or irretrievable commitment of resources prohibited by subsection (d) of this section; and

(B) it establishes such reasonable mitigation and enhancement measures, including, but not limited to, live propagation, transplantation, and habitat acquisition and improvement, as are necessary and appropriate to minimize the adverse effects of the agency action upon the endangered species, threatened species, or critical habitat concerned.

Any final determination by the Committee under this subsection shall be considered final agency action for purposes of chapter 7 of Title 5.

(2) (A) Except as provided in subparagraph (B), an exemption for an agency action granted under paragraph (1) shall constitute a permanent exemption with respect to all endangered or threatened species for the purposes of completing such agency action

(i) regardless whether the species was identified in the biological assessment; and

(ii) only if a biological assessment has been conducted under subsection (c) of this section with respect to such agency action.

(B) An exemption shall be permanent under subparagraph (A) unless

(i) the Secretary finds, based on the best scientific and commercial data available, that such exemption would result in the extinction of a species that was not the subject of consultation under subsection (a) (2) of this section or was not identified in any biological assessment conducted under subsection (c) of this section, and

(ii) the Committee determines within 60 days after the date of the Secretary's finding that the exemption should not be permanent.

If the Secretary makes a finding described in clause (i), the Committee shall meet with respect to the matter within 30 days after the date of the finding.

(i) Review by Secretary of State

Notwithstanding any other provision of this chapter, the Committee shall be prohibited from considering for exemption any application made to it, if the Secretary of State, after a review of the proposed agency action and its potential implications, and after hearing, certifies, in writing, to the Committee within 60 days of any application made under this section that the granting of any such exemption and the carrying out of such action would be in violation of an international treaty obligation or other international obligation of the United States. The Secretary of State shall, at the time of such certification, publish a copy thereof in the Federal Register.

(j) Notwithstanding any other provision of this chapter, the Committee shall grant an exemption for any agency action if the Secretary of Defense finds that such exemption is necessary for reasons of national security.

(k) Special provisions

An exemption decision by the Committee under this section shall not be a major Federal action for purposes of the National Environmental Policy Act of 1969 [42 U.S.C. Section 4321 et seq.]: Provided, That an environmental impact statement which discusses the impacts upon endangered species or threatened species or their critical habitats shall have been previously prepared with respect to any agency action exempted by such order.

(l) Committee order,

(1) If the Committee determines under subsection (h) of this section that an exemption should be granted with respect to any agency action, the Committee shall issue an order granting the exemption and specifying the mitigation and enhancement measures established pursuant to subsection (h) of this section which shall be carried out and paid for by the exemption applicant in implementing the agency action. All necessary mitigation and enhancement measures shall be authorized prior to the implementing of the agency action and funded concurrently with all other project features.

(2) The applicant receiving such exemption shall include the costs of such mitigation and enhancement measures within the overall costs of continuing the proposed action. Notwithstanding the preceding sentence, the costs of such measures shall not be treated as project costs for the purpose of computing benefit-cost or other ratios for the proposed action. Any applicant may request the Secretary to carry out such mitigation and enhancement measures. The costs incurred by the Secretary in carrying out any such measures shall be paid by the applicant receiving the exemption. No later than one year after the granting of an exemption, the exemption applicant shall submit to the Council on Environmental Quality a report describing its compliance with the mitigation and enhancement measures prescribed by this section. Such a report shall be submitted annually until all such mitigation and enhancement measures have been completed. Notice of the public availability of such reports shall be published in the Federal Register by the Council on Environmental Quality.

(m) Notice

The 60-day notice requirement of section 11 (g) of this Act shall not apply with respect to review of any final determination of the Committee under subsection (h) of this section granting an exemption from the requirements of subsection (a) (2) of this section.

(n) Judicial review

Any person, as defined by section 3 (13) of this Act, may obtain judicial review, under chapter 7 of Title 5, of any decision of the Endangered Species Committee under subsection (h) of this section in the United States Court of Appeals for (1) any circuit wherein the agency action concerned will be, or is being, carried out, or (2) in any case in which the agency action will be, or is being, carried out outside of any circuit, the District of Columbia, by filing in such court within 90 days after the date of issuance of the decision, a written petition for review. A copy of such petition shall be transmitted by the clerk of the court to the Committee and the Committee shall file in the court the record in the proceeding, as provided in section 2112, of Title 28. Attorneys designated by the Endangered Species Committee may appear for, and represent the Committee in any action for review under this subsection.

(o) Exemption as providing exception on taking of endangered species

Notwithstanding sections 1533 (d) and 1538 (a) (1) (B) and (C) of this title, sections 1371 and 1372 of this title, or any regulation promulgated to implement any such section

(1) any action for which an exemption is granted under subsection (h) of this section shall not be considered to be a taking of any endangered species or threatened species with respect to any activity which is necessary to carry out such action; and

(2) any taking that is in compliance with the terms and conditions specified in a written statement provided under subsection (b) (4) (iv) of this section shall not be considered to be a prohibited taking of the species concerned.

(p) Exemptions in Presidentially declared disaster areas

In any area which has been declared by the President to be a major disaster area under the Disaster Relief Act of 1974, the President is authorized to make the determinations required by subsections (g) and (h) of this section for any project for the repair or replacement of a public facility substantially as it existed prior to the disaster under section 401 or 402 of the Disaster Relief Act of 1974, and which the President determines (1) is necessary to prevent the recurrence of such a natural disaster and to reduce the potential loss of human life, and (2) to involve an emergency situation which does not allow the ordinary procedures of this section to be followed. Notwithstanding any other provision of this section, the Committee shall accept the determinations of the President under this subsection.

. . .

Section 9. Prohibited acts

(a) General

(1) Except as provided in sections 6 (g) (2) and 10 of this Act, with respect to any endangered species of fish or wildlife listed pursuant to section 4 of this Act it is unlawful for any person subject to the jurisdiction of the United States to -

(A) import any such species into, or export any such species from the United States;

(B) take (Section 3(19) & Section 3(3)) any such species within the United States or the territorial sea of the United States;

(C) take any such species upon the high seas;

(D) possess, sell, deliver, carry, transport, or ship, by any means whatsoever, any such species taken in violation of subparagraphs (B) and (C);

(E) deliver, receive, carry, transport, or ship in interstate or foreign commerce, by any means whatsoever and in the course of a commercial activity, any such species;

(F) sell or offer for sale in interstate or foreign commerce any such species; or

(G) violate any regulation pertaining to such species or to any threatened species of fish or wildlife listed pursuant to section 4 of this Act and promulgated by the Secretary pursuant to authority provided by this chapter.

(2) Except as provided in sections 6 (g)(2) and 10 of this Act, with respect to any endangered species of plants listed pursuant to section 1533 of this title, it is unlawful for any person subject to the jurisdiction of the United States to -

(A) import any such species into, or export any such species from, the United States;

(B) remove and reduce to possession any such species from areas under Federal jurisdiction; maliciously damage or destroy any such species on any such area; or remove, cut, dig up, or damage or destroy any such species on any other area in knowing violation of any law or regulation of any State or in the course of any violation of a State criminal trespass law;

(C) deliver, receive, carry, transport, or ship in interstate or foreign commerce, by any means whatsoever and in the course of a commercial activity, any such species;

(D) sell or offer for sale in interstate or foreign commerce any such species; or

(E) violate any regulation pertaining to such species or to any threatened species of plants listed pursuant to section 4 of this Act and promulgated by the Secretary pursuant to authority provided by this chapter.

(b) Species held in captivity or controlled environment

(1) The provisions of subsections (a) (1) (A) and (a) (1) (G) of this section shall not apply to any fish or wildlife which was held in captivity or in a controlled environment on (A) December 28, 1973, or (B) the date of the publication in the Federal

Register of a final regulation adding such fish or wildlife species to any list published pursuant to subsection (c) of section 4 of this Act: Provided, That such holding and any subsequent holding or use of the fish or wildlife was not in the course of a commercial activity. With respect to any act prohibited by subsections (a) (1) (A) and (a) (1) (G) of this section which occurs after a period of 180 days from (i) December 28, 1973, or (ii) the date of publication in the Federal Register of a final regulation adding such fish or wildlife species to any list published pursuant to subsection (c) of section 4 of this Act, there shall be a rebuttable presumption that the fish or wildlife involved in such act is not entitled to the exemption contained in this subsection.

(2) (A) The provisions of subsection (a) (1) of this section shall not apply to -

(i) any raptor legally held in captivity or in a controlled environment on November 10, 1978; or

(ii) any progeny of any raptor described in clause (i); until such time as any such raptor or progeny is intentionally returned to a wild state.

(B) Any person holding any raptor or progeny described in subparagraph (A) must be able to demonstrate that the raptor or progeny does, in fact, qualify under the provisions of this paragraph, and shall maintain and submit to the Secretary, on request, such inventories, documentation, and records as the Secretary may by regulation require as being reasonably appropriate to carry out the purposes of this paragraph. Such requirements shall not unnecessarily duplicate the requirements of other rules and regulations promulgated by the Secretary.

(c) Violation of Convention

(1) It is unlawful for any person subject to the jurisdiction of the United States to engage in any trade in any specimens contrary to the provisions of the Convention, or to possess any specimens traded contrary to the provisions of the Convention, including the definitions of terms in article I thereof.

(2) Any importation into the United States of fish or wildlife shall, if

(A) such fish or wildlife is not an endangered species listed pursuant to section 4 of this Act but is listed in Appendix II to the Convention,

(B) the taking and exportation of such fish or wildlife is not contrary to the provisions of the Convention and all other applicable requirements of the Convention have been satisfied,

(C) the applicable requirements of subsections (d), (e), and (f) of this section have been satisfied, and

(D) such importation is not made in the course of a commercial activity, be presumed to be an importation not in violation of any provision of this chapter or any regulation issued pursuant to this chapter.

(d) Imports and exports

(1) In general

It is unlawful for any person, without first having obtained permission from the Secretary, to engage in business -

(A) as an importer or exporter of fish or wildlife (other than shellfish and fishery products which (i) are not listed pursuant to section 4 of this Act as endangered species or threatened species, and (ii) are imported for purposes of human or animal consumption or taken in waters under the jurisdiction of the United States or on the high seas for recreational purposes) or plants; or

(B) as an importer or exporter of any amount of raw or worked African elephant ivory.

(2) Requirements

Any person required to obtain permission under paragraph (1) of this subsection shall -

(A) keep such records as will fully and correctly disclose each importation or exportation of fish, wildlife, plants, or African elephant ivory made by him and the subsequent disposition made by him with respect to such fish, wildlife, plants, or ivory;

(B) at all reasonable times upon notice by a duly authorized representative of the Secretary, afford such representative access to his place of business, an opportunity to examine his inventory of imported fish, wildlife, plants, or African elephant ivory and the records required to be kept under subparagraph (A) of this paragraph, and to copy such records; and

(C) file such reports as the Secretary may require.

(3) Regulations

The Secretary shall prescribe such regulations as are necessary and appropriate to carry out the purposes of this subsection.

(4) Restriction on consideration of value or amount of African elephant ivory imported or exported

In granting permission under this subsection for importation or exportation of African elephant ivory, the Secretary shall not vary the requirements for obtaining such permission on the basis of the value or amount of ivory imported or exported under such permission.

(e) REPORTS.

It is unlawful for any person importing or exporting fish or wildlife (other than shellfish and fishery products which

(1) are not listed pursuant to section 4 of this Act as endangered or threatened species, and

(2) are imported for purposes of human or animal consumption or taken in waters under the jurisdiction of the United States or on the high seas for recreational purposes) or plants to fail to file any declaration or report as the Secretary deems

necessary to facilitate enforcement of this Act or to meet the obligations of the Convention.

(f) DESIGNATION OF PORTS.

(1) It is unlawful for any person subject to the jurisdiction of the United States to import into or export from the United States any fish or wildlife (other than shellfish and fishery products which (A) are not listed pursuant to section 4 of this Act as endangered species or threatened species, and (B) are imported for purposes of human or animal consumption or taken in waters under the jurisdiction of the United States or on the high seas for recreational purposes) or plants, except at a port or ports designated by the Secretary of the Interior. For the purposes of facilitating enforcement of this Act and reducing the costs thereof, the Secretary of the Interior, with approval of the Secretary of the Treasury and after notice and opportunity for public hearing, may, by regulation, designate ports and change such designations. The Secretary of the Interior, under such terms and conditions as he may prescribe, may permit the importation or exportation at nondesignated ports in the interest of the health or safety of the fish or wildlife or plants, or for other reasons if, in his discretion, he deems it appropriate and consistent with the purpose of this subsection.

(2) Any port designated by the Secretary of the Interior under the authority of section 4(d) of the Act of December 5, 1969 (16 U.S.C. 666cc 4(d)), shall, if such designation is in effect on the day before the date of the enactment of this Act, be deemed to be a port designated by the Secretary under paragraph (1) of this subsection until such time as the Secretary otherwise provides.

(g) VIOLATIONS.

It is Unlawful for any person subject to the jurisdiction of the United States to attempt to commit, solicit another to commit, or cause to be committed, any offense defined in this section.

EXCEPTIONS

SEC. 10.

(a) PERMITS.

(1) The Secretary may permit, under such terms and conditions as he shall prescribe

(A) any act otherwise prohibited by section 9 for scientific purposes or to enhance the propagation or survival of the affected species, including, but not limited to, acts necessary for the establishment and maintenance of experimental popula-tions pursuant subsection (j); or (B) any taking otherwise prohibited by section 9 (a) (1) (B) if such taking is incidental to, and not the purpose of, the carrying out of an otherwise lawful activity.

(2) (A) No permit may be issued by the Secretary authorizing any taking referred to in paragraph (1) (B) unless the applicant therefor submits to the Secretary a conservation plan that specifies

(i) the impact which will likely result from such taking;

(ii) what steps the applicant will take to minimize and mitigate such impacts, and the funding that will be available to implement such steps;

(iii) what alternative actions to such taking the applicant considered and the reasons why such alternatives are not being utilized; and

(iv) such other measures that the Secretary may require as being necessary or appropriate for purposes of the plan.

(B) If the Secretary finds, after opportunity for public comment, with respect to a permit application and the related conservation plan that

(i) the taking will be incidental;

(ii) the applicant will, to the maximum extent practicable, minimize and mitigate the impacts of such taking;

(iii) the applicant will ensure that adequate funding for the plan will be provided;

(iv) the taking will not appreciably reduce the likelihood of the survival and recovery of the species in the wild; and

(v) the measures, if any, required under subparagraph (A) (iv) will be met; and he has received such other assurances as he may require that the plan will be implemented, the Secretary shall issue the permit. The permit shall contain such terms and conditions as the Secretary deems necessary or appropriate to carry out the purposes of this paragraph, including, but not limited to, such reporting requirements as the Secretary deems necessary for determining whether such terms and conditions are being complied with.

(C) The Secretary shall revoke a permit issued under this paragraph if he finds that the permittee is not complying with the terms and conditions of the permit.

(b) Hardship exemptions

(1) If any person enters into a contract with respect to a species of fish or wildlife or plant before the date of the publication in the Federal Register of notice of consideration of that species as an endangered species and the subsequent listing of that species as an endangered species pursuant to section 4 of this Act will cause undue economic hardship to such person under the contract, the Secretary, in order to minimize such hardship, may exempt such person from the application of section 9 (a) of this Act to the extent the Secretary deems appropriate if such person applies to him for such exemption and includes with such application such information as the Secretary may require to prove such hardship; except that

(A) no such exemption shall be for a duration of more than one year from the date of publication in the Federal Register of notice of consideration of the species concerned, or shall apply to a quantity of fish or wildlife or plants in excess of that specified by the Secretary;

(B) the one-year period for those species of fish or wildlife listed by the Secretary as endangered prior to the effective date of this Act shall expire in accordance with the terms of section 3 of the Act of December 5, 1969 (83 Stat. 275); and

(C) no such exemption may be granted for the importation or exportation of a specimen listed in Appendix I of the Convention which is to be used in a commercial activity.

(2) As used in this subsection, the term "undue economic hardship" shall include, but not be limited to:

(A) substantial economic loss resulting from inability caused by this chapter to perform contracts with respect to species of fish and wildlife entered into prior to the date of publication in the Federal Register of a notice of consideration of such species as an endangered species;

(B) substantial economic loss to persons who, for the year prior to the notice of consideration of such species as an endangered species, derived a substantial portion of their income from the lawful taking of any listed species, which taking would be made unlawful under this chapter; or

(C) curtailment of subsistence taking made unlawful under this chapter by persons (i) not reasonably able to secure other sources of subsistence; and (ii) dependent to a substantial extent upon hunting and fishing for subsistence; and (iii) who must engage in such curtailed taking for subsistence purposes.

(3) The Secretary may make further requirements for a showing of undue economic hardship as he deems fit. Exceptions granted under this section may be limited by the Secretary in his discretion as to time, area, or other factor of applicability.

(c) Notice and review
    The Secretary shall publish notice in the Federal Register of each application for an exemption or permit which is made under this section. Each notice shall invite the submission from interested parties, within thirty days after the date of the notice, of written data, views, or arguments with respect to the application; except that such thirty-day period may be waived by the Secretary in an emergency situation where the health or life of an endangered animal is threatened and no reasonable alternative is available to the applicant, but notice of any such waiver shall be published by the Secretary in the Federal Register within ten days following the issuance of the exemption or permit. Information received by the Secretary as a part of any application shall be available to the public as a matter of public record at every stage of the proceeding.

(d) Permit and exemption policy
    The Secretary may grant exceptions under subsections (a) (1) (A) and (b) of this section only if he finds and publishes his finding in the Federal Register that (1) such exceptions were applied for in good faith, (2) if granted and exercised will not operate to the disadvantage of such endangered species, and (3) will be consistent with the purposes and policy set forth in section 2 of this Act.

(e) Alaska natives

(1) Except as provided in paragraph (4) of this subsection the provisions of this Act shall not apply with respect to the taking of any endangered species or threatened species, or the importation of any such species taken pursuant to this section, by –

(A) any Indian, Aleut, or Eskimo who is an Alaskan Native who resides in Alaska; or

(B) Any non-native permanent resident of an Alaskan native village; if such taking is primarily for subsistence purposes. Non-edible byproducts of species taken pursuant to this section may be sold in interstate commerce when made into authentic native articles of handicrafts and clothing; except that the provisions of this subsection shall not apply to any non-native resident of an Alaskan native village found by the Secretary to be not primarily dependent upon the taking of fish and wildlife for consumption or for the creation and sale of authentic native articles of handicrafts and clothing.

. . .

(j) Experimental populations

(1) For purposes of this subsection, the term "experimental population" means any population (including any offspring arising solely therefrom) authorized by the Secretary for release under paragraph (2), but only when, and at such times as, the population is wholly separate geographically from nonexperimental populations of the same species.

(2)(A) The Secretary may authorize the release (and the related transportation) of any population (including eggs, propagules, or individuals) of an endangered species or a threatened species outside the current range of such species if the Secretary determines that such release will further the conservation of such species.

(B) Before authorizing the release of any population under subparagraph (A), the Secretary shall by regulation identify the population and determine, on the basis of the best available information, whether or not such population is essential to the continued existence of an endangered species or a threatened species.

(C) For the purposes of this chapter, each member of an experimental population shall be treated as a threatened species; except that

(i) solely for purposes of section 7 (other than subsection (a) (1) thereof), an experimental population determined under subparagraph (B) to be not essential to the continued existence of a species shall be treated, except when it occurs in an area within the National Wildlife Refuge System or the National Park System, as a species proposed to be listed under section 4; and

(ii) critical habitat shall not be designated under this chapter for any experimental population determined under subparagraph (B) to be not essential to the continued existence of a species.

(3) The Secretary, with respect to populations of endangered species or threatened species that the Secretary authorized, before October 13, 1982, for release in geographical areas separate from the other populations of such species, shall determine by regulation which of such populations are an experimental population for the purposes of this subsection and whether or not each is essential to the continued existence of an endangered species or a threatened species.

Section 11. Penalties and enforcement

(a) Civil penalties

(1) Any person who knowingly violates, and any person engaged in business as an importer or exporter of fish, wildlife, or plants who violates, any provision of this chapter, or any provision of any permit or certificate issued hereunder, or of any regulation issued in order to implement subsection (a) (1) (A), (B), (C), (D), (E), or (F), (a) (2) (A), (B), (C), or (D), (c), (d) (other than regulation relating to recordkeeping or filing of reports), (f) or (g) of section 9 of this Act, may be assessed a civil penalty by the Secretary of not more than $25,000 for each violation. Any person who knowingly violates, and any person engaged in business as an importer or exporter of fish, wildlife, or plants who violates, any provision of any other regulation issued under this chapter may be assessed a civil penalty by the Secretary of not more than $12,000 for each such violation. Any person who otherwise violates any provision of this Act, or any regulation, permit, or certificate issued hereunder, may be assessed a civil penalty by the Secretary of not more than $500 for each such violation. No penalty may be assessed under this subsection unless such person is given notice and opportunity for a hearing with respect to such violation. Each violation shall be a separate offense. Any such civil penalty may be remitted or mitigated by the Secretary. Upon any failure to pay a penalty assessed under this subsection, the Secretary may request the Attorney General to institute a civil action in a district court of the United States for any district in which such person is found, resides, or transacts business to collect the penalty and such court shall have jurisdiction to hear and decide any such action. The court shall hear such action on the record made before the Secretary and shall sustain his action if it is supported by substantial evidence on the record considered as a whole.

. . .

(b) Criminal violations

(1) Any person who knowingly violates any provision of this chapter, of any permit or certificate issued hereunder, or of any regulation issued in order to implement subsection (a) (1) (A), (B), (C), (D), (E), or (F); (a) (2) (A), (B), (C), or (D), (c), (d) (other than a regulation relating to recordkeeping, or filing of reports), (f), or (g) of section 9 of this Act shall, upon conviction, be fined not more than $50,000 or imprisoned for not more than one year, or both. Any person who knowingly violates any provision of any other regulation issued under this chapter shall, upon conviction, be fined not more than $25,000 or imprisoned for not more than six months, or both.

(2) The head of any Federal agency which has issued a lease, license, permit, or other agreement authorizing a person to import or export fish, wildlife, or plants, or

to operate a quarantine station for imported wildlife, or authorizing the use of Federal lands, including grazing of domestic livestock, to any person who is convicted of a criminal violation of this chapter or any regulation, permit, or certificate issued hereunder may immediately modify, suspend, or revoke each lease, license, permit, or other agreement. The Secretary shall also suspend for a period of up to one year, or cancel, any Federal hunting or fishing permits or stamps issued to any person who is convicted of a criminal violation of any provision of this chapter or any regulation, permit, or certificate issued hereunder. The United States shall not be liable for the payments of any compensation, reimbursement, or damages in connection with the modification, suspension, or revocation of any leases, licenses, permits, stamps, or other agreements pursuant to this section.

. . .

ENDANGERED PLANTS

SEC. 12. The Secretary of the Smithsonian Institution, in conjunction with other affected agencies, is authorized and directed to review (1) species of plants which are now or may become endangered, or threatened and (2) methods of adequately conserving such species, and to report to Congress, within one year after the date of the enactment of this Act, the results of such review including recommendations for new legislation or the amendment of existing legislation.

. . .

References
Endangered Species Act, *U.S. Code*, Vol. 16, secs. 1531–1544. (1973), <http://www.fws.gov/r9endspp/esa.html>, accessed 2 February 2000.
U.S. Fish and Wildlife Service, "Endangered Species Databases, <http://endangered.fws.gov/delist.asc>, accessed 2 February 2000.

HUMAN RESEARCH SUBJECTS

This executive summary is from the report "Adequacy of Federal Protections for Human Subjects in Research," presented by the National Bioethics Advisory Commission on 3 May 1999.

Extending Federal Protections for Human Research Subjects to All Americans, Executive Summary

In 1997, President Clinton stated that "science must respect the dignity of every American. We must never allow our citizens to be unwitting guinea pigs in scientific experiments...." That same month, the National Bioethics Advisory Commission (NBAC) resolved, as a matter of ethical principle, that no person should be enrolled in research without the twin protections of informed consent and independent review of the research. NBAC notes with concern that this goal remains unmet.

In particular, the protections of the Federal Policy for the Protection of Human Subjects, also known as the Common Rule, do not extend to all Americans; the

Common Rule applies only to subjects in research regulated by the Food and Drug Administration (FDA) or to subjects in research sponsored by some Federal departments and agencies.[1] Among the Common Rule's most important protections are the requirements for informed consent by research subjects and for independent review of the research by a local Institutional Review Board (IRB). Despite the fact that many research institutions voluntarily apply the Common Rule—even to their privately financed research—there are other significant sectors of privately funded research that remain ungoverned either by State or Federal law.

NBAC finds that the absence of Federal jurisdiction over much privately funded research means that the U.S. government cannot know how many Americans currently are subjects in experiments, cannot influence how they have been recruited, cannot ensure that research subjects know and understand the risks they are undertaking, and cannot ascertain whether they have been harmed.

Not only does this prevent the Federal Government from protecting Americans enrolling in research, but it affects the Federal Government's ability to craft policies governing emerging technologies. While preparing its 1997 report *Cloning Human Beings,* for example, NBAC noted that the Common Rule's lack of jurisdiction over privately funded research made it impossible to rely on IRBs as the primary mechanism for protecting human subjects against inappropriate uses of those technologies.

Implementation of the Common Rule

Beginning in 1996, Federal departments and agencies responded to NBAC's request for information pursuant to Executive Order 12975. NBAC is pleased to report that agencies have responded to the Executive Order not only by reporting on their current protections, but by evaluating those protections and taking steps to strengthen them. Based on the agency reports and actions and on its own investigations and contracted studies, NBAC concludes that the Common Rule has significantly reduced, but not eliminated, the possibility for harm to human subjects. As a result, NBAC also concludes that there is a need for significant improvement, both to enforce Federal protections and to make their implementation less burdensome for Federal agencies and researchers.

Research regulated by the FDA or sponsored by one of the Federal departments or independent agencies that have adopted the Common Rule requires prior approval and continuing oversight by an IRB. NBAC has found that all the Federal departments and agencies that sponsor substantial amounts of biomedical research with human subjects have implemented these requirements. On the other hand, several departments and agencies that sponsor behavioral and other nonbiomedical research have not fully implemented the provisions of the Common Rule, despite the fact that such research may pose serious nonphysical risks, such as loss of insurance or employment, discrimination, incarceration, and invasion of privacy. Although various Federal regulations do provide protections for certain vulnerable populations, these are not incorporated in the Common Rule. In addition, NBAC notes that the Common Rule does not require any special protections for especially vulnerable populations, such as children.

NBAC has identified occasions when nonbiomedical research that posed more than minimal risk was conducted on certain vulnerable populations; in some of these instances, the research was supported or conducted by one of the agencies that has not adopted additional protections, such as those found in Subparts B, C, or D of the Department of Health and Human Services regulations.

Federal departments and agencies do face obstacles in fully implementing the Common Rule. NBAC notes that many agencies find the Common Rule confusing or its provisions too burdensome in light of the type or amount of research they sponsor. Although some agencies have been taking steps to bring themselves into compliance with the Common Rule, nearly all of them agree that increased protection of human subjects cannot be achieved without additional staffing and highly visible statements of commitment from the leadership of their respective departments. Some also have suggested that a central authority governing human subjects research could help to interpret the Common Rule's requirements, create the oversight structures needed for its implementation, and advise on ethically complex protocols.

NBAC also finds that centralized leadership is needed to achieve consistent interpretation of key statutory and regulatory requirements. Lack of a single authority also means that improvement in human subjects protections, such as those specific to vulnerable populations, requires that every affected department independently adopt new regulations. This is inevitably slower and more inefficient than adoption by a central authority.

For example, in its 1998 report *Research Involving Persons with Mental Disorders That May Affect Decisionmaking Capacity,* NBAC observed that some affected agencies were hard-pressed to reconcile their agency mission of fostering much-needed research into the causes and cures for mental illness with the shared Federal commitment to paying scrupulous attention to the interests of vulnerable human subjects. In addition, the absence of a single, authoritative Federal office to oversee human subjects protections will make it difficult to ensure that all affected departments will issue regulations implementing NBAC's recommendations; indeed, similar recommendations were made 20 years ago by another national bioethics commission regarding the same population, but they were never adopted.

Ensuring Adequate and Accountable Local Oversight

The decentralized local system is sorely strained by inadequate staffing and education of IRBs; by the explosion in research activity; by emerging ethical issues arising from ethical issues raised by epidemiological and public health research; by the trend toward collaborative, multi-centered research; and by an absence of comprehensive public accountability.

NBAC's work highlights many of these problems and offers some solutions. Its upcoming report on research involving human biological materials, for example, suggests some solutions to the difficult problem of applying current Federal protections to epidemiological research on stored tissue, while its project on international research norms is revealing the dilemmas posed by collaborative research across national boundaries. Its report, *Research Involving Persons with Mental Disorders That May Affect Decisionmaking Capacity,* emphasized the need to maintain the public's trust in the integrity of the scientific endeavor. To that end, NBAC suggested that "IRBs can effectively use the mechanisms of audit (both internal and external) and disclosure to improve accountability and inspire public confidence in their oversight activities."

Conclusion

NBAC finds that the current Federal regulations have served to prevent most recurrences of the gross abuses associated with biomedical research in the earlier part of this century. Nonetheless, some abuses still occur, and the system is in need of significant revision in order to provide clear, efficient, and authoritative guidance to Federal departments and agencies and to ensure that local oversight is effective and accountable to the public. This is essential to improving protections for human research subjects. It is also a necessary first step toward extending these protections to those Americans not yet protected by any State or Federal standards for human subjects in research.

Note

1. As of 1997, 16 departments and independent agencies had formally adopted the Common Rule as signatories. In addition, one other independent agency—the Central Intelligence Agency (CIA)—had adopted the Rule in accordance with Executive Order 12333:

Departments

U.S. Department of Agriculture

Department of Commerce

Department of Defense

Department of Education

Department of Energy

Department of Health and Human Services

Department of Housing and Urban Development

Department of Justice

Department of Transportation

Department of Veterans Affairs

Independent Agencies

International Development Cooperation Agency (Agency for International Development)

Consumer Product Safety Commission

Environmental Protection Agency

National Aeronautics and Space Administration

National Science Foundation

Social Security Administration

Central Intelligence Agency*

Office of Science and Technology Policy**

* As of 1997, the CIA was not a formal signatory to the Common Rule but had adopted the Rule in accordance with Executive Order 12333. At the time of this report, the CIA is in the process of becoming a formal signatory.

** The Office of Science and Technology Policy is a signatory to the Common Rule, even though it does not itself conduct or support research directly. It has "accepted" the Common Rule, but does not have its own Code of Federal Regulations.

Reference

"NBAC Summary of Preliminary Findings: Adequacy of Federal Protections for Human Subjects in Research," executive summary, National Bioethics Advisory Commission, <http://bioethics.gov/finalmay3.pdf>.

GENETIC TESTING

This executive summary is from the final report of the Task Force on Genetic Testing, which was created by the National Institutes of Health–Department of Energy Working Group on Ethical, Legal, and Social Implications of Human Genome Research. The report is titled "Promoting Safe and Effective Genetic Testing in the United States," and dated September 1997.

Executive Summary

The rapid pace of discovery of genetic factors in disease has improved our ability to predict risks of disease in asymptomatic individuals. We have learned how to prevent the manifestations of a few of these diseases and treat some others. Gene therapy is being actively investigated.

Despite remarkable progress much remains unknown about the risks and benefits of genetic testing.

- No effective interventions are yet available to improve the outcome of most inherited diseases.
- Negative (normal) test results might not rule out future occurrence of disease.
- Positive test results might not mean the disease will inevitably develop.

It is primarily in the context of their unknown potential risks and benefits that the Task Force considers genetic testing.

Origin and Work of the Task Force

The Task Force was created by the National Institutes of Health (NIH)-Department of Energy (DOE) Working Group on Ethical, Legal, and Social Implications (ELSI) of Human Genome Research to review genetic testing in the United States and make recommendations to ensure the development of safe and effective genetic tests. The Task Force has defined safety and effectiveness to encompass not only the validity and utility of genetic tests, but their delivery in laboratories of assured quality, and their appropriate use by health care providers and consumers.

The Working Group invited organizations with a stake in genetic testing to submit nominations from which it selected members of the Task Force. In addition, the Working Group invited five agencies in the Department of Health and Human Services (HHS) to send nonvoting liaison members to the Task Force. Principles and recommendations of the Task Force appear in bold-faced type.

Definition of Genetic Tests

Genetic test – The analysis of human DNA, RNA, chromosomes, proteins, and certain metabolites in order to detect heritable disease-related genotypes, mutations, phenotypes, or karyotypes for clinical purposes. Such purposes include predicting risk of disease, identifying carriers, and establishing prenatal and clinical diagnosis or prognosis. Prenatal, newborn and carrier screening, as well as testing in high risk families, are included. Tests for metabolites are covered only when they are undertaken with high probability that an excess or deficiency of the metabolite indicates the presence of heritable mutations in single genes. Tests conducted purely for research are excluded from the definition, as are tests for somatic (as opposed to heritable) mutations, and testing for forensic purposes.

The Task Force is primarily concerned about predictive uses of genetic tests performed in healthy or apparently healthy people. Predictive test results do not necessarily mean that the disease will inevitably occur or remain absent; they replace the individual's prior risks based on population data or family history with risks based on genotype. Some, but not all, predictive genetic testing falls under the rubric "genetic screening," a search in a population for persons possessing certain genotypes.

The Need for Recommendations

For the most part, genetic testing in the United States has developed successfully, providing options for avoiding, preventing, and treating inherited disorders. However, problems arise as a result of current practices.

- Sometimes, genetic tests are introduced before they have been demonstrated to be safe, effective, and useful (see chapter 2 and appendices 5 and 6).
- There is no assurance that every laboratory performing genetic tests for clinical purposes meets high standards (see chapter 3).
- Often, the informational materials distributed by academic and commercial genetic testing laboratories do not provide sufficient information to fill in the gaps in providers' and patients' understanding of genetic tests (see appendix 4).
- In the next few years, a greater burden for offering genetic testing will fall on providers who have little formal training or experience in genetics.

In this report, the Task Force does not recommend policies for specific tests but suggests a framework for ensuring that new tests meet criteria for safety and effectiveness before they are unconditionally released, thereby reducing the likelihood of premature clinical use. **The focus of the Task Force on potential problems in no way is intended to detract from the benefits of genetic testing. Its overriding goal is to recommend policies that will reduce the likelihood of damaging effects so the benefits of testing can be fully realized undiluted by harm.**

Need for an Advisory Committee on Genetic Testing

The Task Force calls on the Secretary of Health and Human Services (HHS) to establish an advisory committee on genetic testing in the Office of the Secretary. Members of the committee should represent the stakeholders in genetic testing, including professional societies (general medicine, genetics, pathology, genetic counseling), the biotechnology industry, consumers, and insurers, as well as other interested parties. The various HHS agencies with activities related to the development and delivery of genetic tests should send nonvoting representatives to the advisory committee, which can also coordinate the relevant activities of these agencies and private organizations. The Task Force leaves it to the Secretary to determine the relationship of this advisory committee to others that may be created in the broader area of genetics and public policy, of which genetic testing is only one part.

The committee would advise the Secretary on implementation of recommendations made by the Task Force in this report to ensure that (a) the introduction of new genetic tests into clinical use is based on evidence of their analytical and clinical validity, and utility to those tested; (b) all stages of the genetic testing process in clinical laboratories meet quality standards; (c) health providers who offer and order genetic tests have sufficient competence in genetics and genetic testing to protect the well-being of their patients; and (d) there be continued and expanded availability of tests for rare genetic diseases.

The Task Force recognizes the widely inclusive nature of genetic tests. It is therefore essential that the advisory committee recommend policies for the Secretary's consideration by which agencies and organizations implementing recommendations can determine those genetic tests that need stringent scrutiny. Stringent scrutiny is indicated when a test has the ability to predict future inherited disease in healthy or apparently healthy people, is likely to be used for that purpose, and when no confirmatory test is available. The advisory committee or its designate should define additional indications.

In order to carry out its functions, the advisory committee should have its own staff and budget.

The Task Force further recommends that the Secretary review the accomplishments of the advisory committee on genetic testing after 2 full years of operation and determine whether it should continue to operate.

Overarching Principles. In making recommendations on safety and effectiveness, the Task Force concentrated on test validity and utility, laboratory quality, and provider competence. It recognizes, however, that other issues impinge on testing, and problems may arise from testing. Regarding these issues, the Task Force endorses the following principles.

Informed Consent. The Task Force strongly advocates written informed consent. The failure of the Task Force to comment on informed consent for other uses does not imply that it should not be obtained.

Test Development. **Informed consent for any validation study must be obtained whenever the specimen can be linked to the subject from which it came.**

Testing in Clinical Practice. **(1) It is unacceptable to coerce or intimidate individuals or families regarding their decision about predictive genetic testing. Respect for personal autonomy is paramount. People being offered testing must understand that testing is voluntary. Their informed consent should be obtained. Whatever decision they make, their care should not be jeopardized.**

(2) Prior to the initiation of predictive testing in clinical practice, health care providers must describe the features of the genetic test, including potential consequences, to potential test recipients.

Newborn Screening. **(1) If informed consent is waived for a newborn screening test, the analytical and clinical validity and clinical utility of the test must be established, and parents must be provided with sufficient information to understand the reasons for screening.** By clinical utility, the Task Force means that interventions to improve the outcome of the infant identified by screening have been proven to be safe and effective.

(2) For those disorders for which newborn screening is available but the tests have not been validated or shown to have clinical utility, written parental consent is required prior to testing.

Prenatal and Carrier Testing. **Respect for an individual's/couples' beliefs and values concerning tests undertaken for assisting reproductive decisions is of paramount importance and can best be maintained by a nondirective stance.** One way of ensuring that a nondirective stance is taken and that parents' decisions are autonomous, is through requiring informed consent.

Testing of Children. **Genetic testing of children for adult onset diseases should not be undertaken unless direct medical benefit will accrue to the child and this benefit would be lost by waiting until the child has reached adulthood.**

Confidentiality. Protecting the confidentiality of information is essential for all uses of genetic tests.

(1) Results should be released only to those individuals for whom the test recipient has given consent for information release. Means of transmitting information should be chosen to minimize the likelihood that results will become available to unauthorized persons or organizations. Under no circumstances should results with identifiers be provided to any outside parties, including employers, insurers, or government agencies, without the test recipient's written consent.

(2) Health care providers have an obligation to the person being tested not to inform other family members without the permission of the person tested, except in extreme circumstances.

Discrimination. **No individual should be subjected to unfair discrimination by a third party on the basis of having had a genetic test or receiving an abnormal genetic test result.** Third parties include insurers, employers, and educational and other institutions that routinely inquire about the health of applicants for services or positions.

Consumer Involvement in Policy Making. Although other stakeholders are concerned about protecting consumers, they cannot always provide the perspective brought by consumers themselves, the end users of genetic testing. **Consumers**

should be involved in policy (but not necessarily in technical) decisions regarding the adoption, introduction, and use of new, predictive genetic tests.

References

Holtzman, N., and M. Watson (eds.), "Promoting Safe and Effective Genetic Testing in the United States," executive summary, Task Force on Genetic Testing, September 1997, <http://www.nhgri.nih.gov/ELSI/TFGT_final/index.html>, accessed 2 February 2000.

————, *Promoting Safe and Effective Genetic Testing in the United States:* Final Report of the Task Force on Genetic Testing, Baltimore: Johns Hopkins University Press, 1997.

CLONING

The following introduction and summary are from the report, "Cloning Human Beings," presented by the National Bioethics Advisory Commission in June 1997.

Introduction

The idea that humans might someday be cloned—created from a single somatic cell without sexual reproduction—moved further away from science fiction and closer to a genuine scientific possibility on February 23, 1997. On that date, *The Observer* broke the news that Ian Wilmut, a Scottish scientist, and his colleagues at the Roslin Institute were about to announce the successful cloning of a sheep by a new technique. The technique involved transplanting the genetic material of an adult sheep, apparently obtained from a differentiated somatic cell, into an egg from which[2] the nucleus had been removed. The resulting birth of the sheep, named Dolly, on July 5, 1996 appears to mark yet another milestone in our ability to control, refine, and amplify the forces of nature.

The Scottish sheep experiment was different from prior attempts to create identical offspring from a pair of adult animals. It used a cloning technique to produce an animal that was a genetic twin of an adult sheep. Put another way, Dolly contained the genetic material of only one parent. This technique of transferring a nucleus from a somatic cell into an egg is an extension of research that had been ongoing for over 40 years using nuclei derived from non-human embryonic and fetal cells. The demonstration that nuclei from cells derived from an adult animal could be "reprogrammed," or that the full genetic complement of such a cell could be reactivated well into the chronological life of the cell, is what sets the results of this experiment apart from prior work. In this report the technique, first reported by Wilmut, of nuclear transplantation using nuclei derived from somatic cells other than those of an embryo or fetus is referred to as "somatic cell nuclear transfer."

For some time, scientific evidence has suggested that the genetic material contained in differentiated somatic cells may retain the potential to direct the development of healthy fertile adult animals, but its capacity to do so remained unproved (Di Bernadino 1997). The Roslin experiment, therefore, was a significant scientific event with potentially profound implications since it brings us closer to the possibility of developing a capacity to create clone human beings in an asexual manner. Although for the past ten years scientists have routinely cloned sheep and cows from embryo cells, this was the first successful experiment using the nucleus of a somatic cell from an adult animal to clone an animal that matured to a fully developed state.

The issues surrounding the cloning of human beings have long been the subject of periodic concern and debate among philosophers, scientists, ethicists, and others, particularly following the publication of Joshua Lederberg's 1966 article on cloning in the *American Naturalist* (Lederberg 1966). Nevertheless, the impact of these most recent developments on our national psyche has been quite remarkable. Some commentators have suggested that the furor aroused by the new possibility for cloning is out of proportion to most of the ethical, legal, and moral issues it raises, since these same issues have been raised by previous developments and are simply emerging again in a novel and striking form. Nevertheless, it is important to acknowledge that the possibilities raised by this new technique certainly would be unprecedented and that some would consider its use to be a truly radical step. This type of cloning would involve three novel developments: the replacement of sexual procreation with asexual replication of an existing set of genes; the ability to predetermine the genes of a child; and the ability to create many genetically identical offspring.

Some scientists were surprised that the technical barriers of cell differentiation and development seemingly could be so easily overcome when using somatic cells as the source for nuclear transfer. The public—including many members of the scientific community—responded to Dolly with a combination of fascination, hope for useful new understandings of human biology, and profound concern—even alarm—about the prospect of being able to create whole humans from a single somatic cell via nuclear transfer cloning techniques. Although much of the initial public reaction was one of fear, concern, and serious moral reservations about the potential use or abuse of this new technological capacity, a few voices were heard cautiously suggesting that a better understanding of cell dynamics in humans and animals might enable us to develop new cures for various diseases. Thus, it is important to reflect not only on the dangers and ethical reservations but also on the potential human benefits from the use of this type of cloning that might arise in such areas as treating particular infertility problems, transplanting cells or tissues, or preventing certain genetically transmitted harms to offspring.

A few of the initial objections to this new type of cloning were either speculative or based on simple misunderstandings, such as, that cloning would allow for the instantaneous creation of a fully grown adult from the cells of an individual. Other fears stemmed from the incorrect idea that an exact copy, although much younger, of an existing person could be made. This fear reflects an erroneous belief that one's genes bear a simple relationship to the physical and psychological traits that make up a person. Although genes provide the building blocks for each individual, it is the interaction among a person's genetic inheritance, the physical and cultural environment, and the process of learning that result in the uniqueness of each individual human. Thus, the idea that nuclear transplantation cloning could be used to re-create exemplary or evil people has no scientific basis and is simply false.

Other objections to nuclear transplantation cloning, however, are based on carefully articulated philosophical ideals, deep cultural commitments, or religious beliefs, and these deserve continuing and careful consideration. These objections reflect deeply held beliefs about the value of human individuality and personal autonomy, the meaning of family and the value of a child, respect for human life and the natural world, and the preservation of the integrity of the human species.

Many public leaders in the United States responded to the announcement about Dolly with immediate and strong condemnation of any attempt to clone human beings in this new manner. The reasons ranged from frightening science fiction imagery to the judgment that cloning of human beings is a serious violation of basic human rights and human dignity. The reaction abroad was similar, with many nations seemingly ready—indirectly or directly—to prohibit cloning human beings in this fashion. Indeed, many international organizations such as UNESCO and the Council of Europe have a long-established and well-articulated concern that research and clinical applications in biology and genetics remain consistent with a fundamental commitment to human dignity and human rights. To date, at least Argentina, Australia, Great Britain, Denmark, Germany, and Spain have enacted laws banning cloning human beings. Unfortunately, some of the deep concerns supporting such views and associated legislation are stated in vague or overly broad terms. The widespread public discomfort, even revulsion, about cloning human beings deserves the best articulation possible, a task that takes time and requires the considered reflections of diverse groups within American society and abroad.

Within days of the published report of the apparently successful cloning of a sheep in this new manner, President Clinton instituted a ban on federal funding for research related to cloning of human beings. In addition, the President asked the recently appointed National Bioethics Advisory Commission (NBAC) to address within ninety days the ethical and legal issues that surround the subject of cloning human beings. This provided a welcome opportunity for initiating a thoughtful analysis of the many dimensions of the issue, including a careful consideration of the potential risks and benefits. It also presented an occasion to review the current legal status of cloning and the potential constitutional challenges that might be raised if new legislation were enacted to restrict the creation of a child through somatic cell nuclear transfer.

The Commission began its discussions fully recognizing that any effort in humans to transfer a somatic cell nucleus into an enucleated egg involves the creation of an embryo, with the apparent potential to be implanted in utero and developed to term. Ethical concerns surrounding issues of embryo research have recently received extensive analysis and deliberation in our country. Indeed, federal funding for human embryo research is severely restricted, although there are few restrictions on human embryo research carried out in the private sector. Thus, under current law, the use of somatic cell nuclear transfer to create an embryo solely for research purposes is already restricted in cases involving federal funds. There are, however, no current regulations on the use of private funds for this purpose.

The unique prospect, vividly raised by Dolly, is the creation of a new individual genetically identical to an existing (or previously existing) person—a "delayed" genetic twin. This prospect has been the source of the overwhelming public concern about such cloning. While the creation of embryos for research purposes alone always raises serious ethical questions, the use of somatic cell nuclear transfer to create embryos raises no new issues in this respect. The unique and distinctive ethical issues raised by the use of somatic cell nuclear transfer to create children relate to, for example, serious safety concerns, individuality, family integrity, and treating children as objects. Consequently, the Commission focused its attention on the use of such techniques for the purpose of creating an embryo which would then be implanted in a woman's uterus and brought to term. It also expanded its analysis of this particular issue to encompass activities in both the public and private sector.

Note

2. A somatic cell is any cell of the embryo, fetus, child, or adult which contains a full complement of two sets of chromosomes; in contrast with a germ cell, i.e., an egg or a sperm, which contains only one set of chromosomes.

Controlling Nature

Humankind's efforts to control nature date back as far as recorded history. In particular, domesticated plants and animals have been the mainstay of our agricultural heritage. Over time human mastery over nature often has been met, quite understandably, with opposition and concern, and frequently has been considered by some to be an affront to the natural order of things or by others to be at odds with interpretations of God's revealed word. Indeed many myths and legends, ancient as well as modern, deal directly with humankind's on-going struggle to ensure that the benefits of our new technological capacities clearly outweigh the harms—both expected and unexpected. The idea that our growing technological mastery is filled with moral ambiguity and capable of both vast good and catastrophic evil is deeply embedded in many cultural traditions.

A prime example is the mythology of the Argo, the first ship, in classical Greek culture. The Greeks see the initial act of shipbuilding as both the origin of culture and the origin of decline. While sailing enables one to encounter other persons and other possibilities, it also brings marauders and war, and its very existence bespeaks the danger of unlimited human desire. Thus, the ability to build and sail boats is both a boon and a curse. Euripides' *Medea* starts with a lament about the trees that were cut down to build the Argo and the other troubles that followed:

> Would that the Argo had never winged its way to the land of Colchis. . . .
> Would that pine trees had never been felled in the glens of Mount
> Pelion and furnished oars for the hands of the heroes who at Pelias'
> command set forth in quest of the Golden Fleece.

Concern about our tools and technology has been greatly accelerated with the coming of modern industrialized societies. Is it possible, some now wonder, that our confidence in human competence and technology may be just another myth? How, some are now asking, can we find some moral compass or moral limit to our desire to master everything and possess all? Only such limits, many would say, can save us from the moral ambiguity of our own cleverness.

In recent years, concern about humankind's control over nature has been particularly acute in relation to the new moral choices created by the stunning developments in the biomedical sciences, especially in the area of human reproduction. Although personal reproductive health is considered to be, in most cases, a private matter, ongoing controversies regarding the moral standing of human genetic material and particular human interventions in procreation have focused public attention on the ethical and legal implications of new reproductive techniques. In many cases, initial fears give way to cautious acceptance, but a wariness lingers that is easily reawakened with each new advance.

Artificial insemination by donor, for example, was considered a form of adultery when first introduced in the 1940s. It is now a widely used and accepted practice in the treatment of infertility, although some continue to have serious reservations. When prenatal diagnosis was introduced in the late 1960s, the public simultaneously welcomed the opportunity to prevent lethal disease in newborns but worried about the use of such techniques to select "vanity" characteristics or nonmedical traits in offspring. The birth of Louise Brown, conceived via in vitro fertilization, in 1978 was another dramatic event, providing a new and controversial means to parenthood. With all of these technical advances, there has been a continuing debate about safety, legality, ethical acceptability, and the government's right to intervene in private matters.

Research itself, not just its clinical application, has often sparked debate. For example, research involving human fetuses has been a subject of intense national debate and disagreement for over two decades (Institute of Medicine, 1994). Federal research in this area continues to be restricted to that which has potential therapeutic benefit to the fetus, or involves no more than minimum risk to the fetus even if potential benefit to the mother can be demonstrated. Restrictions also remain regarding embryo research. Despite the recommendations of the National Institutes of Health Human Embryo Research Panel (1994), that certain targeted and carefully regulated research using early human embryos be eligible for federal funds, in December 1994 the President directed NIH not to allocate federal funds for research programs that involved the creation of human embryos solely for research purposes. This issue was also addressed by Congress, which inserted language in the FY96 and FY97 appropriations bills that widened the presidential ban to prohibit virtually all human embryo research conducted with federal funds. Work in this area continues in the United States, but it is largely limited to the private sector, and thus takes place without any federal regulation.

Recombinant DNA research represents another example of controversy and intense debate. In the 1970s, concerns about the safety of unintended release of recombinant organisms led to a voluntary research moratorium in the scientific community and the development of guidelines (Fredrickson, 1991). Similarly, all experiments involving gene therapy (treatment of specific diseases by inserting human genes into human patients) are subject to review and approval by a federal body.

As segments of human DNA or human cells became the focus of study and the objects of manipulation, their use as research materials raised increasingly important ethical issues about how these materials are obtained, transformed, and, in some cases, used to develop commercial products (Office of Technology Assessment, 1987). Such research with human genetic material generates questions about respect for persons and the human body, and the value and moral status to be placed on cells and tissues.

Genetic and reproductive technologies also cause concern because of the specter of eugenics and of real or imagined social control through manipulation of human genes. Genetic control suggests broken taboos, and, in the words of Henry David Thoreau, implies that "men have become the tools of their tools"(Blank, 1981). While these concerns are often set against and partly attributable to a backdrop of fiction, fantasy, and misunderstanding, they are, more importantly, related to profound concerns regarding the nature of humankind and its relationship to other aspects of the natural world.[3] When the bizarre and fantastic scenarios are removed, we are left with a myriad of reactions: sincere expressions of opposition; serious moral concerns; new hope for a better understanding of human biology and the prospect of combating currently untreatable afflictions; calls for more study; and guarded statements about the need for some measure of control (Macklin, 1994; 1997).

Note

3. With respect to interesting fiction consider Aldous Huxley's *Brave New World* (1932), David Rorvik's unsubstantiated claim of successful human cloning in *In His Image* (1978), and popular films such as *The Boys from Brazil* (1978) and *Jurassic Park* (1993) in which cloning leads to dire, doomsday consequences.

Controlling Science

With some notable exceptions, the scientific community has enjoyed for centuries a great deal of autonomy in directing and regulating its research agenda. Since mid-century, however, demands for external regulation have increased, in part because much research, particularly in the biological sciences, is publicly funded and therefore requires some additional measure of accountability. More importantly, society has become more sensitive to concerns about the dangers—particularly to human participants—of the research itself and its future consequences. Thus, our evolving moral sensibilities together with the spectacular advances in biomedical science have generated new ethical concerns. As Bernard Davis of Harvard Medical School and others have noted, society sometimes seeks to regulate or restrict research when it poses the specters of dangerous or unfamiliar products, powers, or ideas (Davis, 1980).

The regulation of science has thus become part of the landscape, particularly for those who receive federal funds (Office of Technology Assessment, 1986). In addition to environmental, health, occupational, and safety regulations, scientists must also comply with animal welfare and human subjects protections and abide by restrictions and moratoria on specific types of research. Because science is both a public and social enterprise and its application can have profound impact, society recognizes that the freedom of scientific inquiry is not an absolute right and scientists are expected to conduct their research according to widely held ethical principles. There are times when limits on scientific freedom must be imposed, even if such limits are perceived as an impediment by an individual scientist. Moreover, appropriate ethical constraints are a matter for both scientists and the broader public to formulate and implement. At the same time, limits on freedom of inquiry must be justified, and impositions on such freedom should satisfy certain conditions—for example, that the limits are not arbitrary, that they emerge from the thoughtful balancing of costs and benefits, that they are not unnecessarily oppressive, that they do not lightly impinge on long established rights and freedoms, that there is some continuing public discourse with those affected by the ban, and that such limitations be open to reconsideration in the light of new information and new understanding.

Consideration of Ethical and Religious Perspectives

When the President asked NBAC to take up the issue of the cloning of human beings he admonished that "any discovery that touches upon human creation is not simply a matter of scientific inquiry, it is a matter of morality and spirituality as well." Although well aware that the United States Constitution prohibits the establishment of policies that are solely motivated by religious beliefs, NBAC shared the President's concern and sought out testimony about the cloning of human beings from leading scholars from a variety of religious traditions. In the same spirit NBAC also commissioned a background paper on the positions a number of religious traditions have taken or are considering on the cloning of human beings.

NBAC felt this was especially important because religious traditions influence and shape the moral views of many U.S. citizens and religious teachings over the centuries have provided an important source of ideas and inspiration. Although in a pluralistic society particular religious views cannot be determinative for public policy decisions that bind everyone, policy makers should understand and show respect for diverse moral ideas regarding the acceptability of cloning of human beings in this new manner.

Although some religious responses to the cloning of human beings through somatic cell nuclear transfer are tied tightly to particular scriptural texts or other faith commitments, often these ideas can be stated forcefully in terms understandable and persuasive to all persons, irrespective of specific religious beliefs. For example, appeal may be made to a view of human nature or of human reason, rather than exclusively to a religious source of knowledge such as scripture or revelation.

NBAC also wanted to determine whether various religious traditions, despite their distinctive sources of authority and argumentation, reach similar conclusions about this type of human cloning. A convergence of views across these traditions, as well as across secular traditions, would be instructive, even if not necessarily determinative, for public policy. While many Americans look to their religious faiths for moral guidance on issues, other sources of moral knowledge and insight are also important. Many moral considerations that would be widely acknowledged as legitimate do not depend for their force on particular religious commitments or a specific philosophical outlook. For example, the conviction that it is wrong to harm a child is broadly shared among Americans. If you inquire why it is wrong to harm a child, people may give different answers. Some may refer to their religious convictions that a child is a gift from God. Others may say that it is always wrong to harm an innocent person without some compelling reason. To many people, this is a bedrock principle of ethics, even if it has no single, universally acknowledged foundation in a specific religious or philosophical tradition. Rather, it finds its foundation in many different understandings of morality, some religious, some secular. Moral ideas such as the obligation not to inflict harm on others are accessible to all Americans and, therefore, can provide a robust foundation for public policy.

America has a vibrant tradition of ethical dialogue in which all are invited to participate. What moral considerations deserve our attention and which are the most important in responding to a particular issue? These are questions that arise with every new controversy. Whether one's ethical beliefs come from theological commitments, philosophical arguments, or from hard-won life experience, all voices should be welcome to the conversation, and all thoughtful views are entitled to a respectful hearing. While tolerance is a widely accepted virtue in America, it is important to remind ourselves that it is built on the idea of mutual respect and the capacity to accept, whenever possible, the moral worth of others with whom one may disagree. Tolerance, therefore, means both agreeing to disagree and accepting the challenge of sustaining a community where moral authority will, to some extent, always be contested.

Policy makers, therefore, need to consider a range of moral views when they try to determine whether a particular policy is ethically justifiable as well as politically feasible. A particular policy may not be politically feasible, for instance, if it evokes thoughtful, widespread and vigorous moral opposition. In such circumstances its social costs may outweigh its putative benefits, and additional education and deliberation may be required before new policies are put in place.

Consideration of Law and Public Policy

The public policy chosen with respect to the cloning of human beings via somatic cell nuclear transfer should reflect a keen knowledge of the science, our best judgments about the ethics of attempting such an experiment, and our traditions regarding limitations on individual actions in the name of the common good. Americans in this era, relative to earlier generations, have a wide interest in and substantial knowledge of science. Nevertheless, in the weeks following the report of Dolly, the public, the media, and even some scientists demonstrated a surprising lack of understanding of the science involved in cloning. NBAC believes that public debate about issues such as human cloning requires an even more educated populace. Science policy has become public policy, which can be decided wisely only by an informed nation.

American tradition has been to avoid prohibiting or regulating personal activities, absent a compelling reason related to effects on others or society as a whole. Where the individual actions are expressions of fundamental rights, such as the right to free speech or the right to privacy, the reasons for limitation must be compelling, and the limitations made as minimal as possible. The possibility of cloning human beings in this new fashion appears to raise concerns about direct physical harms to the children who may result. This in itself is sufficient to justify a prohibition on such attempts at this time, even if such efforts were to be character-ized as the exercise of a fundamental right to procreate. More speculative psycho-logical harms to the child, and effects on the moral, religious, and cultural values of society may be enough to justify continued prohibitions in the future, but more time is needed for discussion and evaluation of these concerns.

In its discussion of potential policy options, NBAC considered the relative benefits of achieving an immediate prohibition through federal legislation on cloning human beings using somatic cell nuclear transfer techniques. It also considered more indirect means to deter such experiments.

Indirect, non-legislative options considered by NBAC include cooperation by the private sector, both research and clinical, in a moratorium on such experiments and/or clinical practice, and the continued prohibition of the use of federal funds to support such experiments. The American Medical Association, the World Medical Association, and the World Health Organization, for example, have already called for such a moratorium on clinical activities.

NBAC also weighed, in terms of nuclear transplantation cloning, the potential impact of a possible legislative measure to extend basic human subjects protections to all research conducted in the United States. This would insure that any research efforts to clone a human in this manner would, along with all other research using human subjects, be covered by the twin protections of informed consent and appropriate scientific review to insure an ethically acceptable balance between risks and benefits. In light of the early state of animal research in this area, such protections should prevent such cloning research from going forward at this time.

Finally, NBAC recognized that cooperation with other governments in the enforcement of any common elements of our respective policies could strengthen any of the measures adopted by the United States. Because science is a global endeavor, international cooperation would ensure consistency across borders and enhance public confidence in scientific research generally.

Process of NBAC and Organization of the Report

The results of NBAC's 90-day analysis are presented in this report. In its deliberations, NBAC focused its discussion on the science of the cloning of human beings using the somatic cell nuclear transfer technique, and the ethical, religious, legal, and regulatory implications of cloning human beings in this manner. To aid in these tasks NBAC invited testimony from an array of scientists, scientific societies, ethicists, theologians, and legal experts, and heard from a wide variety of interested parties during the public comment session at each meeting. In addition, it commissioned numerous background papers from recognized experts to inform its work.

This report consists of five chapters in addition to this one. Chapter Two describes the scientific developments that preceded and made possible the cloning of Dolly and speculates on potential applications of this and related technologies. Chapter Three presents some of the key themes in religious interpretations and evaluations of human cloning. Chapter Four outlines the numerous ethical concerns raised by the prospect of cloning human beings via somatic cell nuclear transfer. Chapter Five discusses the legal and policy issues considered by the NBAC as it pondered various recommendations. The final section, Chapter Six, presents the recommendations made by NBAC in response to the President's request.

In many instances, NBAC found itself moving at a rapid pace in only partly charted waters. In those times it relied on its individual and collective wisdom, judgment, and moral foundations, and the advice of others. NBAC argued and debated the issues as it searched for appropriate formulations of the problem and for the wisdom to suggest useful policy options. While the members of NBAC learned a great deal during its deliberations, we could not reach a resolution on all of the issues before us. Nevertheless, it was able to accomplish two things. First, it developed a set of recommendations, which are set out in Chapter Six. Second, it agreed that it was important to take a number of steps to ensure the continuation of an informed national discussion of these issues and other developments in the biomedical sciences and clinical practices that have an impact on our moral lives and cultural traditions.

Executive Summary

The idea that humans might someday be cloned—created from a single somatic cell without sexual reproduction—moved further away from science fiction and closer to a genuine scientific possibility on February 23, 1997. On that date, *The Observer* broke the news that Ian Wilmut, a Scottish scientist, and his colleagues at the Roslin Institute were about to announce the successful cloning of a sheep by a new technique which had never before been fully successful in mammals. The technique involved transplanting the genetic material of an adult sheep, apparently obtained from a differentiated somatic cell, into an egg from which the nucleus had been removed. The resulting birth of the sheep, named Dolly, on July 5, 1996, was different from prior attempts to create identical offspring since Dolly contained the genetic material of only one parent, and was, therefore, a "delayed" genetic twin of a single adult sheep.

This cloning technique is an extension of research that had been ongoing for over 40 years using nuclei derived from non-human embryonic and fetal cells. The demonstration that nuclei from cells derived from an adult animal could be "reprogrammed," or that the full genetic complement of such a cell could be reactivated well into the chronological life of the cell, is what sets the results of this experiment apart from prior work. In this report the technique, first described by Wilmut, of nuclear transplantation using nuclei derived from somatic cells other than those of an embryo or fetus is referred to as "somatic cell nuclear transfer."

Within days of the published report of Dolly, President Clinton instituted a ban on federal funding related to attempts to clone human beings in this manner. In addition, the President asked the recently appointed National Bioethics Advisory Commission (NBAC) to address within ninety days the ethical and legal issues that surround the subject of cloning human beings. This provided a welcome opportunity for initiating a thoughtful analysis of the many dimensions of the issue, including a careful consideration of the potential risks and benefits. It also presented an occasion to review the current legal status of cloning and the potential constitutional challenges that might be raised if new legislation were enacted to restrict the creation of a child through somatic cell nuclear transfer cloning.

The Commission began its discussions fully recognizing that any effort in humans to transfer a somatic cell nucleus into an enucleated egg involves the creation of an embryo, with the apparent potential to be implanted in utero and developed to term. Ethical concerns surrounding issues of embryo research have recently received extensive analysis and deliberation in the United States. Indeed, federal funding for human embryo research is severely restricted, although there are few restrictions on human embryo research carried out in the private sector. Thus, under current law, the use of somatic cell nuclear transfer to create an embryo solely for research purposes is already restricted in cases involving federal funds. There are, however, no current federal regulations on the use of private funds for this purpose.

The unique prospect, vividly raised by Dolly, is the creation of a new individual genetically identical to an existing (or previously existing) person—a "delayed" genetic twin. This prospect has been the source of the overwhelming public concern about such cloning. While the creation of embryos for research purposes alone always raises serious ethical questions, the use of somatic cell nuclear transfer to create embryos raises no new issues in this respect. The unique and distinctive ethical issues raised by the use of somatic cell nuclear transfer to create children relate to, for example, serious safety concerns, individuality, family integrity, and treating children as objects. Consequently, the Commission focused its attention on the use of such techniques for the purpose of creating an embryo which would then be implanted in a woman's uterus and brought to term. It also expanded its analysis of this particular issue to encompass activities in both the public and private sector.

In its deliberations, NBAC reviewed the scientific developments which preceded the Roslin announcement, as well as those likely to follow in its path. It also considered the many moral concerns raised by the possibility that this technique could be used to clone human beings. Much of the initial reaction to this possibility was negative. Careful assessment of that response revealed fears about harms to the children who may be created in this manner, particularly psychological harms associated with a possibly diminished sense of individuality and personal autonomy. Others expressed concern about a degradation in the quality of parenting and family life.

In addition to concerns about specific harms to children, people have frequently expressed fears that the widespread practice of somatic cell nuclear transfer cloning would undermine important social values by opening the door to a form of eugenics or by tempting some to manipulate others as if they were objects instead of persons. Arrayed against these concerns are other important social values, such as protecting the widest possible sphere of personal choice, particularly in matters pertaining to procreation and child rearing, maintaining privacy and the freedom of scientific inquiry, and encouraging the possible development of new biomedical breakthroughs.

To arrive at its recommendations concerning the use of somatic cell nuclear transfer techniques to create children, NBAC also examined long-standing religious traditions that guide many citizens' responses to new technologies and found that religious positions on human cloning are pluralistic in their premises, modes of argument, and conclusions. Some religious thinkers argue that the use of somatic cell nuclear transfer cloning to create a child would be intrinsically immoral and thus could never be morally justified. Other religious thinkers contend that human cloning to create a child could be morally justified under some circumstances, but hold that it should be strictly regulated in order to prevent abuses.

The public policies recommended with respect to the creation of a child using somatic cell nuclear transfer reflect the Commission's best judgments about both the ethics of attempting such an experiment and its view of traditions regarding limitations on individual actions in the name of the common good. At present, the use of this technique to create a child would be a premature experiment that would expose the fetus and the developing child to unacceptable risks. This in itself might be sufficient to justify a prohibition on cloning human beings at this time, even if such efforts were to be characterized as the exercise of a fundamental right to attempt to procreate.

Beyond the issue of the safety of the procedure, however, NBAC found that concerns relating to the potential psychological harms to children and effects on the moral, religious, and cultural values of society merited further reflection and deliberation. Whether upon such further deliberation our nation will conclude that the use of cloning techniques to create children should be allowed or permanently banned is, for the moment, an open question. Time is an ally in this regard, allowing for the accrual of further data from animal experimentation, enabling an assessment of the prospective safety and efficacy of the procedure in humans, as well as granting a period of fuller national debate on ethical and social concerns. The Commission therefore concluded that there should be imposed a period of time in which no attempt is made to create a child using somatic cell nuclear transfer.[1]

Within this overall framework the Commission came to the following conclusions and recommendations:

I. The Commission concludes that at this time it is morally unacceptable for anyone in the public or private sector, whether in a research or clinical setting, to attempt to create a child using somatic cell nuclear transfer cloning. The Commission reached a consensus on this point because current scientific information indicates that this technique is not safe to use in humans at this point. Indeed, the Commission believes it would violate important ethical obligations were clinicians or researchers to attempt to create a child using these particular technologies, which are likely to involve unacceptable risks to the fetus and/or potential child. Moreover, in addition to safety concerns, many other serious ethical concerns have been identified, which require much more widespread and careful public deliberation before this technology may be used.

The Commission, therefore, recommends the following for immediate action:

A continuation of the current moratorium on the use of federal funding in support of any attempt to create a child by somatic cell nuclear transfer.

An immediate request to all firms, clinicians, investigators, and professional societies in the private and non-federally funded sectors to comply voluntarily with the intent of the federal moratorium. Professional and scientific societies should make clear that any attempt to create a child by somatic cell nuclear transfer and implantation into a woman's body would at this time be an irresponsible, unethical, and unprofessional act.

II. The Commission further recommends that:

Federal legislation should be enacted to prohibit anyone from attempting, whether in a research or clinical setting, to create a child through somatic cell nuclear transfer cloning. It is critical, however, that such legislation include a sunset clause to ensure that Congress will review the issue after a specified time period (three to five years) in order to decide whether the prohibition continues to be needed. If state legislation is enacted, it should also contain such a sunset provision. Any such legislation or associated regulation also ought to require that at some point prior to the expiration of the sunset period, an appropriate oversight body will evaluate and report on the current status of somatic cell nuclear transfer technology and on the ethical and social issues that its potential use to create human beings would raise in light of public understandings at that time.

III. The Commission also concludes that:

Any regulatory or legislative actions undertaken to effect the foregoing prohibition on creating a child by somatic cell nuclear transfer should be carefully written so as not to interfere with other important areas of scientific research. In particular, no new regulations are required regarding the cloning of human DNA sequences and cell lines, since neither activity raises the scientific and ethical issues that arise from the attempt to create children through somatic cell nuclear transfer, and these fields of research have already provided important scientific and biomedical advances. Likewise, research on cloning animals by somatic cell nuclear transfer does not raise the issues implicated in attempting to use this technique for human cloning, and its continuation should only be subject to existing regulations regarding the humane use of animals and review by institution-based animal protection committees.

If a legislative ban is not enacted, or if a legislative ban is ever lifted, clinical use of somatic cell nuclear transfer techniques to create a child should be preceded by research trials that are governed by the twin protections of independent review and informed consent, consistent with existing norms of human subjects protection.

The United States Government should cooperate with other nations and international organizations to enforce any common aspects of their respective policies on the cloning of human beings.

IV. The Commission also concludes that different ethical and religious perspectives and traditions are divided on many of the important moral issues that surround any attempt to create a child using somatic cell nuclear transfer techniques. Therefore, the Commission recommends that:

The federal government, and all interested and concerned parties, encourage widespread and continuing deliberation on these issues in order to further our understanding of the ethical and social implications of this technology and to enable society to produce appropriate long-term policies regarding this technology should the time come when present concerns about safety have been addressed.

V. Finally, because scientific knowledge is essential for all citizens to participate in a full and informed fashion in the governance of our complex society, the Commission recommends that:

Federal departments and agencies concerned with science should cooperate in seeking out and supporting opportunities to provide information and education to the public in the area of genetics, and on other developments in the biomedical sciences, especially where these affect important cultural practices, values, and beliefs.

Note

1. The Commission also observes that the use of any other technique to create a child genetically identical to an existing (or previously existing) individual would raise many, if not all, of the same non-safety-related ethical concerns raised by the creation of a child by somatic cell nuclear transfer.

References

National Bioethics Advisory Commission, "Cloning Human Beings," Introduction, <http://bioethics.gov/pubs/cloning1/chapter1.pdf>, June 1997.

————, "Cloning Human Beings: Report and Recommendations of the National Bioethics Advisory Commission," Rockville, MD, June 1997.

STEM CELL RESEARCH

New reports of stem-cell research led to the following exchange between U.S. President Bill Clinton and Harold Shapiro, the chair of the National Bioethics Advisory Commission. The president's letter is dated 14 November 1998, and Shapiro's response is dated 20 November 1998. Another letter dated 4 March 1999 and signed by three dozen Nobel laureates added their point of view: "Stem cell research has enormous potential for the effective treatment of human disease. There is, therefore, a moral imperative to pursue it."

Letter from President Bill Clinton to Harold Shapiro, Chair of the National Bioethics Advisory Commission, 14 November 1998

Dear Dr. Shapiro:

This week's report of the creation of an embryonic stem cell that is part human and part cow raises the most serious of ethical, medical, and legal concerns. I am deeply troubled by this news of experiments involving the mingling of human and non-human species. I am therefore requesting that the National Bioethics Advisory Commission consider the implications of such research at your meeting next week, and to report back to me as soon as possible.

I recognize, however, that other kinds of stem cell research raise different ethical issues, while promising significant medical benefits. Four years ago, I issued a ban on the use of federal funds to create human embryos solely for research purposes; the ban was later broadened by Congress to prohibit any embryo research in the public sector. At that time, the benefits of human stem cell research were hypothetical, while the ethical concerns were immediate. Although the ethical issues have not diminished, it now appears that this research may have real potential for treating such devastating illnesses as cancer, heart disease, diabetes, and Parkinson's disease. With this in mind, I am also requesting that the Commission undertake a thorough review of the issues associated with such human stem cell research, balancing all ethical and medical considerations.

I look forward to receiving your reports on these important issues.

Sincerely,

Bill Clinton

Response by Harold Shapiro, Chair of the National Bioethics Advisory Commission, to President Bill Clinton, 20 November 1998

Dear Mr. President:

I am responding to your letter of November 14, 1998 requesting that the National Bioethics Advisory Commission discuss at its meeting in Miami this week the ethical, medical, and legal concerns arising from the fusion of a human cell with a cow egg.

The Commission shares your view that this development raises important ethical and potentially controversial issues that need to be considered, including concerns about crossing species boundaries and exercising excessive control over nature, which need further careful discussion. This is especially the case if the product resulting from the fusion of a human cell and the egg from a non-human animal is transferred into a woman's uterus and, in a different manner, if the fusion products are embryos even if no attempt is made to bring them to term. In particular, we believe that any attempt to create a child through the fusion of a human cell and a non-human egg would raise profound ethical concerns and should not be permitted.

We devoted time at our meeting to discussing various aspects of this issue, benefiting not only from the expertise of the Commissioners, but from our consultation (via telephone) with Dr. Ralph Brinster, a recognized expert in the field of embryology, from the University of Pennsylvania. Also in attendance at our meeting was Dr. Michael West, of Advanced Cell Technology, who was given an opportunity to answer questions from Commission members. As you know, however, the design and results of this experiment are not yet publicly available, and as a consequence the Commission was unable to evaluate fully its implications.

As a framework for our initial discussion, we found it helpful to consider three questions:

1. Can the product of fusing a human cell with the egg of a non-human animal, if transferred into a woman's uterus, develop into a child?

At this time, there is insufficient scientific evidence to answer this question. What little evidence exists, based on other fusions of non-human eggs with non-human cells from a different species, suggests that a pregnancy cannot be maintained. If it were possible, however, for a child to develop from these fused cells, then profound ethical issues would be raised. An attempt to develop a child from these fused cells should not be permitted.

This objection is consistent with our views expressed in Cloning Human Beings, in which we concluded that

"...(A)t this time it is morally unacceptable for anyone in the public or private sector, whether in a research or clinical setting to attempt to create a child using somatic cell nuclear transfer cloning."

2. Does the fusion of a human cell and an egg from a non-human animal result in a human embryo?

The common understanding of a human embryo includes at least the concept of an organism at its earliest stage of development, which has the potential if transferred to a uterus, to develop in the normal course of events into a living human being. At this time, however, there is insufficient scientific evidence to be able to say whether the combining of a human cell and the egg of a non-human animal results in an embryo in this sense. In our opinion, if this combination does result in an embryo, important ethical concerns arise, as is the case with all research involving human embryos. These concerns will be made more complex and controversial by the fact that these hybrid cells will contain both human and non-human biological material.

It is worth noting that these hybrid cells should not be confused with human embryonic stem cells. Human embryonic stem cells, while derived from embryos, are not themselves capable of developing into children. The use of human embryonic stem cells, for example to generate cells for transplantation, does not directly raise the same type of moral concerns.

3. If the fusion of a human cell and the egg of a non-human animal does not result in an embryo with the potential to develop into a child, what ethical issues remain?

If this line of research does not give rise to human embryos, we do not believe that totally new ethical issues arise. We note that scientists routinely conduct non-controversial and highly beneficial research that involves combining material from human and other species. This research has led to such useful therapies as blood clotting factor for hemophilia, insulin for diabetes, erythropoietin for anemia, and heart valves for transplants. Combining human cells with non-human eggs might possibly lead some day to methods to overcome transplant rejections without the need to create human embryos, or to subject women to invasive, risky medical procedures to obtain human eggs.

We recognize that some of the issues raised by this type of research may also be pertinent to stem cell research in general. We intend to address these and other issues in the report that you requested regarding human stem cell research.

Sincerely,

Harold T. Shapiro
Chair

Reference

"ASCB Stem Cell Research Position," *The ASCB Newsletter*, April 1999: 1–3. (Letter exchange between the President of the United States and the Chair of the National Bioethics Advisory Commission), November 1998, National Bioethics Advisory Commission Web site, <http://bioethics.gov/>, accessed 2 February 2000.

CHAPTER NINE
Career Information

C areers for biologists, and particularly for biologists with doctorates, are branching out from their former stronghold in academia into industry, non-governmental agencies, and government laboratories. Some new graduates view the shift with excitement, while others lament the decline in the quantity of academic positions relative to the number of individuals seeking them.

In the last two decades, the percentage of scientists with Ph.D.s who have been able to land traditional teaching and research positions at universities has dwindled. Of those who are pursuing academic appointments, some are finding it necessary to take postdoctoral appointments for several years before finding tenure-track faculty positions.

This chapter describes the careers available in biology and some of the trends in the current job market. See also Data and Statistics, Chapter 10.

MANY OPPORTUNITIES

A scan of the employment section of a 1999 issue of *Science* magazine, which covers a wide variety of scientific fields, indicates that biologists are in great demand. Most of the biology jobs listed in this prestigious journal, however, require applicants to have a doctoral degree and many require a research background and/or postdoctoral experience.

A Snapshot of the Possibilities

To show the range of jobs available, along with their duties and requirements, this section provides a sample from the many job announcements posted in the 21 May 1999 issue of *Science*. In this issue, as with most others, the ads came mainly from universities or colleges, or from private firms, particularly pharmaceutical companies. Ads described positions available, and requirements for applicants, as follows:

- Two tenure-track assistant/associate professor positions in the department of anatomy, physiology, and pharmacology at a veterinary medicine college; doctorate or equivalent and postdoctoral experience were required; duties included teaching histology or pharmacology, and developing an independent research program.

- Instructor/assistant professor in neurobiology in a university's cell biology and anatomy department; doctorate or equivalent, a background in developmental biology of mice and birds, plus postdoctoral experience were required; duties included teaching and research.

- Temporary microbiology instructor at a university; doctorate and related teaching experience were required; the job entailed teaching.

- Tenure-track assistant professor of food science and technology at a university; a Ph.D. degree was required, and research abstracts were requested.

- Program director for grants administration at the U.S. Department of Agriculture Small Business Innovation Research program; advanced biology degree was required, with a Ph.D. or equivalent experience, and expertise in "biologically oriented agricultural engineering" preferred.

- Protein biochemist at a pharmaceutical research institute; bachelor's or master's degree in biochemistry, at least two years of related experience, plus proficiency in various laboratory techniques were required; duties included performing purification work for protein structural studies.

- Immunologist/immunotoxicologist at a pharmaceutical company; doctorate in immunology or a relevant field, two to five years of related experience, plus expertise in immunochemistry and associated lab techniques were required; duties included managing "scientific and technical activities to assess potential risks of immunologically active novel therapeutics in preclinical drug development."

- Senior research associate at a start-up company involved with genetic technology; master of science degree or equivalent, plus at least two years of experience in the biotechnology industry were required; duties involved work on standard and novel genetic-technology methods.
- Three-year scientist/cell biologist position at a crop-production and plant-protection company; three to five years of experience in a postdoctoral position or as a research scientist, plus experience in project management were preferred; duties included establishing a cell biology group with an emphasis on a particular type of grain.
- Virology scientist at a biotechnology company; Ph.D. in biology or relevant area, at least one year of industry experience or three years of related academic work, plus experience performing specific lab work were required, and educational background in molecular/cellular virology was preferred; duties included the development of "new capacities for molecular detection of viruses and other adventitious agents," and the use of quantitative and qualitative assays.
- Associate director for earth and environmental sciences at a national laboratory; U.S. citizenship was the only requirement listed; duties included responsibility for and to a staff of more than 250 research-and-development employees involved in a number of environmental projects.
- Director of conservation science international programs for a large nature conservancy; Ph.D. in conservation biology, natural resource management, ecosystem ecology, or related field was required; duties involved the management of scientists, specialists, and program managers, and the identification of potential sites for biodiversity work.
- Chief of the salmon analysis branch of a national fisheries service; fishery scientists/ecologist credentials were required; duties included research and the management of a team of researchers who were studying the conservation and restoration of endangered salmon.

The job postings, which appear in every week in *Science*, usually also include dozens of postdoctoral positions for a spectrum of fields. The postings often list very specific duties needed to contribute to a researcher's ongoing project, and applicants may be required to have experience with particular lab techniques or field methods. Some of the ads were broad in their requirements, but the majority sought applicants for very specific projects.

Members of the American Association for the Advancement of Science, which publishes *Science*, can review the listings first-hand in the magazine. Both members and non-members can access the job postings through the association's Web site at <http://aaas.org/>.

Students and beginning professionals can also find information about general academic job openings and a variety of other available positions in the weekly *Chronicle of Higher Education*. This newspaper contains a profusion of job advertisements and provides indices that break down the postings by state and by field. The *Chronicle's* Web site, <http://chronicle.com/>, offers views of the current job listings to subscribers and the previous week's advertisements to non-subscribers.

Other good resources for job openings are journals, newsletters and Web sites that are closely related to the student's area of study. Some examples are described in Chapters 11 and 12. Electronic discussion groups are also excellent places to learn about job opportunities. While some discussion groups focus wholly on job postings, others are mainly research discussion groups with occasional job postings from participants. The latter usually offer a good selection to those students who are seeking temporary research positions that might help build their experience levels in certain areas.

A Broadening Field

As the face of biology changes—the field of astrobiology is off to a quick start (see Chapter 2), and molecular biology and genetics are flourishing (see Chapter 5)—the career opportunities are adjusting to the current needs. The numbers of students in the various biological sciences programs are also conforming, with fewer in some of the traditional courses of study and more in the "hot" fields. In many cases, however, those numbers are deceiving, as some of the students enrolled in "hot" programs are applying their techniques to traditional programs. An example is the marriage between genetics and botany/horticulture, which is yielding a great deal of research on genetically engineered plants.

Interdisciplinary Research

Many current jobs in the biological sciences are centered around research that covers several fields within biology. This interdisciplinary work may even span fields well outside the boundaries of biological sciences, such as physics, chemistry, and engineering. Many granting agencies are even requiring—or strongly recommending—that grant submissions have interdisciplinary components.

The advantage of interdisciplinary work is that scientists with a wide range of viewpoints and expertise can build upon one another's ideas and perform multifaceted work that would not be possible if a single researcher were attempting the same project. The development of transgenic crops is a good example. Both a geneticist and a botanist or a horticulturist can provide vital insights to genetic engineering experiments that neither alone could.

Interdisciplinary efforts also may provide added avenues for applying the research to real-world problems. Pharmaceutical research teams, for example, often include chemists and biologists working side by side, with the chemists making pharmaceuticals to match the biologists' specifications.

As the term "interdisciplinary" has become more firmly ingrained in the research world, biologists and other scientists are tearing down the long-standing barriers between fields and even within fields. Scientists are seeing first-hand the benefits of shared work, not only in results, but in the funds made available to them. Biology graduates with open minds and the willingness to share ideas with scientists outside their discipline will find the shift to interdisciplinary work a stimulating opportunity.

New Fields

One of the newest fields in the biological sciences is astrobiology, which focuses on extraterrestrial life. This field is currently centering much of its attention on Mars and the moons of Jupiter as potential sites of life. The National Aeronautics and Space Administration (NASA) initiated its Astrobiology Institute in 1998, and at about the same time, the University of Washington–Seattle admitted the initial students into what it believes is the world's first doctoral program in the field.

Genomic sequencing has become almost a field in itself. The technique, which allows scientists to unravel an organism's genetic code, has spread to many, many labs as they race to decipher genomes, including the human genetic code. While numerous academic labs are busy deciphering genomes, a growing number of start-up companies have thrown their hats into the ring. A scan of advertisements for biology graduates provides evidence that skills in genetic sequencing are highly valued.

Another area swift on the heels of genomic sequencing is drug development. Information about the genetic sequence of organisms may help guide the development of medicines to counteract disease. Genetic testing, which could reveal a person's predisposition to a certain disease or the likelihood that a fetus carries a genetic mutation, is also on the increase. Gene therapy is becoming a booming field. Researchers are clamoring to find ways to cure disease by altering genes within cells. Careers in all of these areas are thriving.

Neuroscience is seen as an emerging field with opportunities for both basic and applied research. Advanced imaging methods, including such tools as positron emission tomography (PET), are allowing researchers to follow and sort out chemical changes in the brain. The researchers hope to use the findings to help patients with brain disorders, such as epilepsy. Insight about how the brain works is also providing clues to such medical conditions as Alzheimer's disease, multiple sclerosis, and various forms of addiction.

Bioinformatics is another hot field. This field combines computer technology with biological sciences for a myriad of research projects. Besides genetic sequencing, bioinformatics can be extremely helpful in modeling various physiological processes. Researchers, for example, are now feeding detailed information into computer programs and then modeling how certain chemicals affect the physiology of target tissue. This powerful field is playing an important role in drug development.

The emerging field of evolutionary developmental biology, or "evo-devo" as it is known among its practitioners, combines evolutionary biology with developmental biology. In an odd twist, this interdisciplinary field is drawing individual researchers who are interdisciplinary themselves. Those who enlist in evo-devo have backgrounds in both fields of biology, usually beginning in one, then switching to the other later during their graduate work. The combined field brings together the concepts of the evolutionary biologists with the meticulous data collection abilities of the developmental biologists. The field is so new that both positions and research funding are scarce, although interest among many biologists is high.

References

"Director of NASA Astrobiology Institute" (advertisement), *Science*, 9 October 1998: 356.

"Ph.D. for ET?" *Science*, 9 October 1998: 211.

Postdocs

With the shift in funding and the tightened job market in academia has come a similar shift in employment opportunities in the past decade.

Among new doctoral biology graduates, fewer are finding jobs quickly after graduation. In 1982, for example, about 68 percent had landed a job soon after graduation. That contrasts to only about half in 1992.

Additional surveys conducted by the National Research Council showed that the percentage of life-sciences doctoral graduates who were able to obtain faculty positions had changed significantly from 1973 to 1995. For both years, graduates from 2 years, 6 years, and 10 years earlier indicated whether they were able to find faculty positions. In 1973, 40 percent of the 2-year group held faculty positions, compared with just 14 percent in 1995. The numbers for the 6-year group were 57 percent in 1973, but 30 percent in 1995; and for the 10-year group, more than 60 percent in 1973, but fewer than 40 percent in 1995. Even a decade after graduation, the difference remained. More than 60 percent of 1973 graduates held faculty positions, but fewer than 40 percent of 1995 graduates were similarly employed.

Those declining percentages, however, are partially counterbalanced by an increase in the percent of life sciences doctoral scientists holding jobs in industry. In 1973, the figure was 12 percent. In 1995, it jumped to 23 percent.

Many Ph.D. graduates in the 1990s found that their paths to academic, government, or industry positions meant side trips to lengthy (and sometimes multiple) postdoctoral positions. (A postdoctoral position is usually a low-paid, temporary research appointment in someone else's lab, whether it be a professor's or a company's.) Smaller research labs claim to receive about one postdoc application per week, with the larger research institutes reporting hundreds of applications for every permanent position they post.

The number of life sciences doctoral graduates who were still holding a postdoctoral appointment up to two years after graduation rose from 21 percent in 1973 to 53 percent in 1995. Those with postdoctoral appointments up to four years after graduation amounted to 6 percent in 1973, but 29 percent in 1995. The figures for longer-term postdoctoral appointments also showed a disparity: 2 percent of Ph.D.s in 1973 were still in postdoctoral appointments five to six years after graduation, compared with 14 percent of Ph.D.s in 1995.

The positive side to the increase in the length of postdoctoral appointments is that the doctoral graduate is remaining employed in the field and garnering additional research experience, even though the job market for permanent positions is tight. The negative side, as mentioned above, is that postdoctoral appointments are typically low-paying, temporary positions.

In summary, the road to academia is a bit longer than it once was, but it is by no means a dead end.

Reference
Trends in the Early Careers of Life Scientists, 1998, Committee on Dimensions, Causes and Implications of Recent Trends in Careers of Life Scientists, National Research Council.

Preparing for the Job Market

One of the best ways to develop any career is to prepare. The biological sciences are no different. Jobs in academia and industry often are similar in the ways students can prepare for them, but some differences exist.

Tactics for Academia

Most academic positions require that the applicant have a solid research grounding and good publications. In this regard, graduates should think quality over quantity. One publication in a major journal like *Cell* or *Nature* is worth much more than a handful of publications in several small journals that are not peer-reviewed.

The choice of a doctoral adviser can also make a difference. Many universities consider a student's "pedigree," which can be defined as the university that granted the degree and the researcher under whom the student studied. An adviser who is well known and respected among his or her peers can be an asset to a graduate. In other words, it pays to know the right people. A graduate may also find that help from the adviser can open doors.

In addition, doctoral students sometimes find it helpful to build relationships with students and professors at their universities and at other schools while they are still working on their doctoral dissertations. The tactic is called networking in business circles, but biology students often find it can work for them, too. By networking, students can meet like-minded researchers and other students at conferences, in electronic discussion groups, by e-mail exchanges, or through conventional methods, like telephone calls or letters. While "what you know" is important, "who you know" can also be vital to landing an academic position. Upon graduation, students can alert their networks that they are looking for a position and can seek suggestions or guidance. The students may then learn about positions that aren't nationally advertised, or they may receive a few extra references.

Once a university or college requests an interview, the graduate should be prepared to discuss his or her research concisely and in an informative and interesting way. Institutions often require each job applicant to present a colloquium on his or her work. The audience usually includes other faculty members and sometimes graduate students. Even the most qualified and highly recommended applicant can lose standing in a job competition if the colloquium is ill-prepared or poorly presented.

Preparation is the key. A good school, solid publications, a helpful adviser, a strong network, and an effective colloquium presentation can all shorten the run on the postdoctoral treadmill.

Tactics for Industry

Many of the tactics for landing a job in industry parallel those used for academia, but a few extra methods apply.

Graduates may find they have a little more latitude in selling themselves when approaching industry for a position. Because some jobs are in small, start-up companies, the primary and sometimes the only person involved in the hiring process may be the main researcher. In this case, a telephone call can have much greater influence than it would on one member of a university hiring board.

Applicants should also be prepared to make a formal presentation, particularly if they are seeking jobs at larger companies. As in seeking an academic appointment, preparation is critical.

Companies often look for applicants who have an understanding of how the business world works and where the company's emphasis lies. For instance, if the company is trying to apply its research, the applicant should go into the interview with that in mind, and answer questions with an applied—rather than a basic-science—point of view. The attitude that applied research is somehow "beneath" basic research is quickly diminishing as many veteran professors and newer doctoral graduates are beginning to appreciate the positive side of applied work: seeing their work put to use.

In addition, companies may ask applicants about their experience working on teams or in an interdisciplinary research group, which may be the way the company performs its research. Students with and without team experience should consider how to answer those questions ahead of time.

Finally, applicants should be prepared to discuss not just their research projects, but how they can help the company. A review of the company's annual report or Web site can usually provide clues to the company's various emphases.

References

Holden, C., "Science Careers: Playing to Win," *Science*, 23 September 1994: 1905.

Radetsky, P., "The Modern Postdoc: Prepping for the Job Market," *Science*, 23 September 1994: 1909–10.

Stone, R., "Climbing the Industry Career Ladder," *Science*, 23 September 1994: 1911.

Alternative Careers

Academia and industry are not the only employers of biology majors. Biology jobs run the gamut from naturalist to geneticist, from evolutionary biologist to ecosystem manager, and from virologist to entomologist. Even the variety of career paths within just one field of biology can be immense. For example, botany students may find positions in plant nurseries, may run entire botanical gardens, or may become involved in crop management. Others may take positions at natural history museums and collect or classify plants. A few might make a living by working for a nonprofit conservation organization and reviewing sites for potential preservation, and others might work at a university studying the evolution of plants. These are just a few examples within one speciality.

Teaching at the elementary and secondary school levels provides fulfillment for many biology majors. Those interested in this path, however, should consult with the state where they hope to work, to learn about teaching requirements and certificates, and to ensure that they have all of the necessary qualifications.

Students involved in research that has money-making applications, such as genetic engineering or the many offshoots of cellular biology, might even consider beginning a company. This option involves many risks, but those whose companies can survive the growing pains of the first few years usually report that the initial hard work was well worth the effort.

Academic positions outside of research and teaching are often suitable for biology majors. A position in a university's research office can be challenging to a graduate who is interested in many aspects of science and science policy. Graduates with grant-drafting experience or abilities might enjoy positions as grant writers for nonprofit organizations, such as a nature conservancy. Universities and their individual researchers often hire grant writers to assist them with paperwork and report-writing. Some even hire writers to assist with the drafting of research articles for publications. In those cases, the researcher provides the data and usually an outline of the information to be covered, and the writer drafts the article.

In addition, some companies hire scientists to help write articles for their annual reports and newsletters. Scientists who can write for less technically oriented audiences, and who can accept and learn from sometimes-severe criticism and/or editing of their work, will fare well in these positions.

An imaginative biology major can think up dozens of possibilities, but even the most unimaginative can usually think of one or two alternatives to their primary choice for a career. As biology jobs change, and opportunities shift, flexibility is becoming a more and more important asset to a biology graduate.

CHAPTER TEN
Data and Statistics

This chapter contains a variety of charts that provide a snapshot of the field of biological sciences as it relates to education and careers. The data includes recent information, along with comparisons over several years or decades.

EDUCATION

The biological sciences field has encountered a shift in emphasis as new discoveries have been made, new technologies have been developed, and new questions have been formulated. While the number of students in biological sciences appeared to decline from the 1970s to the 1990s, many students from other fields are now studying biology and conducting biological research. A number of the examples of research in this book are taken from the labs of scientists outside biological sciences proper. Nonetheless, the students listing "biological sciences" as their major still make up a large portion of the students in the higher education system.

Within the biological sciences, too, a change has occurred. For example, a study published by the National Center for Education Statistics indicated that from the early 1970s to the early 1990s, the field has shifted in its distribution among its specialty areas. The center divided the field into three broad categories: (a) general biology; (b) microbiology and bacteriology; and (c) zoology, entomology, pathology, pharmacology, and physiology. In the 1970s, biology bachelor's, master's, and doctoral students tended toward the zoology complement much more heavily than the 1990s students. About 16 percent of the bachelor's, master's, and doctoral students—or 8,361 of the total 51,592 students—in these three categories were zoology students in 1973–74, compared with fewer than 10 percent—or 4,761 of the total 48,358 students—in 1993–94. The microbiology and bacteriology contingent also felt a decrease from 6.2 to 5.5 percent of biology students during the two-decade period.

Other reports show similar changes in the number of biological sciences students enrolling in different subject areas within the field. Data published by the National Center for Education Statistics provided more detail for shifts among graduate students and those with postdoctoral appointments. The study indicated that from fall 1986 to fall 1996, some fields, such as cell biology, biometry and epidemiology, pharmacology, and genetics were booming, while others, such as the "staples" of zoology and botany, were lagging. For instance, cell biology and biometry/epide-

Graduate Enrollment in Biological Life Science Fields from Fall 1986 to Fall 1996

	Fall 1986	Fall 1991	Fall 1996
All Sciences and Engineering, Total	415,557	471,262	494,526
Biological Sciences, Total	46,765	51,778	58,127
Anatomy	973	1,051	1,111
Biochemistry	4,873	5,201	5,265
Biology	12,678	13,292	14,635
Biometry/Epidemiology	1,434	2,032	3,001
Biophysics	547	697	833
Botany	3,123	2,694	2,504
Cell Biology	1,716	2,809	3,896
Ecology	1,022	1,180	1,615
Entomology/Parasitology	1,306	1,171	1,234
Genetics	1,262	1,520	1,729
Microbiology	4,371	4,928	4,963
Nutrition	4,259	4,164	4,859
Pathology	1,323	1,449	1,656
Pharmacology	2,078	2,432	2,652
Physiology	2,220	2,332	2,377
Zoology	2,075	2,191	1,910
Other Biosciences	1,505	2,635	3,887

Source: "Table 209—Graduate Enrollment in Science and Engineering Programs in Institutions of Higher Education, by Field of Study: United States and Outlying Areas, 1981 to 1992," Digest of Education Statistics, 1996, National Center for Education Statistics, 1996.

miology more than doubled as the choice of students from fall 1986 to fall 1996, pharmacology jumped nearly 28 percent, and genetics increased by 37 percent. At the same time, those who opted to study zoology dropped by 8 percent and botany fell by 20 percent. The changes in zoology and botany might not be true reflections of the students' interests in the fields, however, because many students who are interested in the two subjects are conducting genetic or other studies of them and may be enrolled under another subject area.

This shift in biology majors' subject areas occurred at a time when more and more students were taking biology courses at the high school level, and the traditional college student was becoming older. Many commuter colleges and universities had been noticing an older student body on average for several years, and the traditional colleges and universities are witnessing the "graying" of their student population. Enrollment totals for 1995–96 showed that more than one quarter of the students enrolled at two-year institutions were 25 or older. Among graduate students, more than three-quarters were 25 or older, and of those, more than one in 10 was more than 35 years old.

Reference
"Table 276–Earned Degrees in Agriculture and Natural Resources Conferred by Institutions of Higher Education, by Level of Degree and Sex of Student: 1970–71 to 1995–96," Chapter 3, *Digest of Education Statistics, 1998*, National Center for Education Statistics, 1998.

EMPLOYMENT

Biological sciences students are finding employment in a variety of occupations as the private sector is opening up. Perhaps the greatest shift in the employment of biological sciences graduates is occurring at the doctoral degree level. The traditional academic jobs of years past have become more scarce, but opportunities in industry appear to be picking up most of the slack. The result is a wider selection of job opportunities.

Data for bachelor's and master's degree recipients are available for the science and engineering fields in general, and for graduates of the life sciences specifically. Reports from the National Science Foundation show that a bachelor's degree in life sciences does not guarantee placement in a life sciences career. A survey of 1994 life sciences graduates showed that nearly 1 in 10 of those who sought employment were unemployed, and more than 1 in 7 of those who were employed held positions not related to their degree.

Employment Success of 1994 Bachelor's Degree Recipients

Major Field	Total Recipients	Employed	Unemployed*	Not in Labor Force
All Science and Engineering Fields	349,700	291,500	16,800	41,400
Life and Related Sciences, Total	62,500	44,700	4,000	13,900
Agricultural and Food Sciences	6,300	5,600	S	S
Biological Sciences	52,500	35,700	3,500	13,200
Environmental Life Sciences, Including Forestry Sciences	3,800	3,300	S	S

Notes

* Those who were not working on April 15 and who were seeking work or were on layoff from a job.

S Data with weighted values less than 100 or unweighted sample sizes of less than 20 are suppressed for reasons of respondent confidentiality and/or data reliability.

Source: "Number of 1994 Science and Engineering Bachelor's Degree Recipients Who Are Employed, Unemployed and Not in the Labor Force, by Field of Degree: April 1995," Chart B-39, National Science Foundation, Division of Science Resources Studies, *Characteristics of Recent Science and Engineering Graduates: 1995*, Detailed Statistical Tables, NSF 97-333, by John Tsapogas, Mary Collins, Lucinda Gray, and Debora Kraft (Arlington, VA, 1997), or see <http://www.nsf.gov/sbe/srs/nsf97333/start.htm>.

Employment Rates and Salaries of 1993 Bachelor's Degree Recipients, by Race

Major Field	Total Recipients (Number)	Primary Education and Employment Status (Number)		Not Full-time Student		Median Salary for Full-Time Employed*
		Full-Time Student	Employed in Science and Engineering	Employed in Other Occupation	Not Employed and Not Full-Time Student	
All Science and Engineering Fields	348,900	82,000	67,900	180,700	18,300	$26,000
Life and Related Sciences						
White, non-Hispanic	46,600	17,300	3,700	23,700	1,900	$23,000
Black, non-Hispanic	2,700	1,300	S	1,300	S	$23,500
Hispanic	3,000	900	S	1,200	S	$23,000
Asian or Pacific Islander	5,900	3,000	S	S	S	S
American Indian/Alaskan Native	400	S	S	200	S	$29,000

Notes

Survey conducted April 1995.

* Salary data are not included for the following groups: self-employed persons, full-time students, and people whose principal job was less than 35 hours per week. Salary data are for principal jobs only.

S Data with weighted values less than 100 or unweighted sample sizes of less than 20 are suppressed for reasons of respondent confidentiality and/or data reliability.

Source: "Number of 1993 Science and Engineering Bachelor's Degree Recipients, by Primary Status, Median Salary, Race/Ethnicity, and Field of Degree: April 1995," Chart S-3, National Science Foundation, Division of Science Resources Studies, *Characteristics of Recent Science and Engineering Graduates: 1995*, Detailed Statistical Tables, NSF 97-333, by John Tsapogas, Mary Collins, Lucinda Gray, and Debora Kraft (Arlington, VA, 1997), or see <http://www.nsf.gov/sbe/srs/nsf97333/start.htm>.

Employment Rates and Salaries of 1993 Bachelor's Degree Recipients, by Gender

| | Total Recipients | Primary Education and Employment Status | | | | |
| | | | Not Full-Time Student | | | |
		Full-Time Student	Employed in Science and Engineering	Employed in Other Occupation	Not Employed and not Full-Time Student	Median Salary for Full-Time Employed*
Life Sciences						
Male	28,100	11,100	2,500	13,900	S	$23,500
Female	30,500	11,400	2,800	14,600	1,700	$23,700

Notes

Survey conducted April 1995.

* Salary data are not included for the following groups: self-employed persons, full-time students, and people whose principal job was less than 35 hours per week. Salary data are for principal jobs only.

S Data with weighted values less than 100 or unweighted sample sizes of less than 20 are suppressed for reasons of respondent confidentiality and/or data reliability.

Source: "Number of 1993 Science and Engineering Bachelor's Degree Recipients, by Primary Status, Median Salary, Sex, and Field of Degree: April 1995," Chart S-2, National Science Foundation, Division of Science Resources Studies, *Characteristics of Recent Science and Engineering Graduates: 1995*, Detailed Statistical Tables, NSF 97-333, by John Tsapogas, Mary Collins, Lucinda Gray, and Debora Kraft (Arlington, VA, 1997), or see <http://www.nsf.gov/sbe/srs/nsf97333/start.htm>.

Master's degree recipients fared better than those who held only bachelor's degrees in obtaining jobs in the life sciences. Part of the reason for their success likely revolves around their ability to obtain graduate teaching or research assistantships at their universities. Students with such assistantships usually receive a stipend and a tuition waiver. In a survey of 1994 master's degree recipients, about 75 percent of those trained in life and related sciences found jobs closely related to their field of study, 16 percent found positions somewhat related, and about 11 percent held jobs not related to their degrees.

Surveys of master's degree recipients also show the diversity of their employment. While nearly half of those who received degrees in 1994 were employed by a university or other educational institution, 35 percent held jobs in private companies. These figures showed a marked difference between a 1994 master's degree recipient in a life sciences field and the average 1994 master's degree recipient in combined sciences and engineering fields. When considering all science and engineering master's degree recipients, 32 percent held jobs in academic settings, while 47 percent were employees of private companies.

Success of Employed 1994 Master's Degree Recipients in Finding Degree-Related Jobs

| | | Relationship of Degree to Job | | |
Major Field	Total Employed	Closely Related	Somewhat Related	Not Related
All Science and Engineering Fields	63,900	44,100	14,700	5,100
Life and Related Sciences, Total	5,500	4,100	900	600
Agricultural and Food Sciences	1,000	700	S	S
Biological Sciences	3,700	2,800	500	S
Environmental Life Sciences Incl. Forestry Sciences	900	600	S	S

Note
S Data with weighted values less than 100 or unweighted sample sizes of less than 20 are suppressed
   for reasons of respondent confidentiality and/or data reliability.

Source: "Number of Employed 1994 Science and Engineering Master's Degree Recipients Having Job Closely, and Somewhat and Not Related to Degree, by Field of Degree: April 1995," Chart B-101, National Science Foundation, Division of Science Resources Studies, *Characteristics of Recent Science and Engineering Graduates: 1995*, Detailed Statistical Tables, NSF 97-333, by John Tsapogas, Mary Collins, Lucinda Gray, and Debora Kraft (Arlington, VA, 1997), or see <http://www.nsf.gov/sbe/srs/nsf97333/start.htm>.

Success in Obtaining Career-Path Jobs among 1994 Master's Degree Recipients

Major Field	Total Recipients	Number Having a Career-Path Job			Number not Having Career-Path Job	Number without Career-Path Job Who Are Seeking One		
		Total	Male	Female		Total	Male	Female
All Science and Engineering Fields	73,400	48,700	30,300	18,400	24,800	9,200	5,800	3,500
Life and Related Sciences, Total	7,400	4,200	1,900	2,300	3,200	600	300	S
Agricultural and Food Sciences	1,200	700	400	S	500	S	S	S
Biological Sciences	5,300	2,800	1,000	1,800	2,600	S	S	S
Environmental Life Sciences Including Forestry Sciences	900	700	S	S	S	S	S	S

Note

S Data with weighted values less than 100 or unweighted sample sizes of less than 20 are suppressed for reasons of respondent confidentiality and/or data reliability.

Source: "Number of 1994 Science and Engineering Master's Degree Recipients Who Have Had a Career-Path Job Since Being Awarded Most Recent Degree, and Number Not Having a Career-Path Job Who Are Seeking One, by Field of Degree: April 1995," Chart B-100, National Science Foundation, Division of Science Resources Studies, *Characteristics of Recent Science and Engineering Graduates: 1995*, Detailed Statistical Tables, NSF 97-333, by John Tsapogas, Mary Collins, Lucinda Gray, and Debora Kraft (Arlington, VA, 1997), or see <http://www.nsf.gov/sbe/srs/nsf97333/start.htm>.

The changes in career opportunities for recipients of doctorates have been the subject of a great deal of discussion. A doctorate once was a ticket to an academic position, but that has changed. Now Ph.D. recipients are spending more time in postdoctoral fellowships before finding academic positions, or they are finding jobs outside of academia. A 1995 National Science Foundation survey of doctorate-holding scientists and engineers showed that while the vast majority held jobs, more than 4 percent were unable to find a position in their fields, despite a desire to do so. In the life sciences, the percentage of those who were employed involuntarily outside of their fields ranged from 3.3 percent in the biological and health sciences to 5.6 percent in the environmental sciences.

Of those doctorate-holding scientists and engineers who obtained positions at institutions of higher learning, 35 percent held full professorships, 22 percent were associate professors, and 19 percent had positions as assistant professors. The remainder held other positions, such as instructor or adjunct faculty member. The percentages in the life sciences include 31.6 percent employed as full professors, 20.5 percent as associate professors, and 20.4 percent as assistant professors.

Selected Employment Characteristics, by Field of Doctorate: 1995

Field of Doctorate	Unemployment Rate (Percent)	Involuntary Out-of-Field Area (Percent)	Labor Force Participation Rate (Percent)
All Sciences and Engineering, Total	1.5	4.2	90.7
Life and Related Sciences	1.7	3.4	90.0
Agricultural and Food Sciences	1.7	3.4	88.0
Biological and Health Sciences	1.7	3.3	90.4
Environmental Sciences	1.8	5.6	86.3

Note: All numbers in the table are estimates derived from a sample.

Source: "Selected Employment Characteristics of Doctoral Scientists and Engineers, by Field of Doctorate: 1995," Table 7, National Science Foundation, Division of Science Resources Studies, *Characteristics of Doctoral Scientists and Engineers in the United States 1995*, NSF 97-319, R. Keith Wilkinson (Arlington, VA, 1997).

Employment in Universities and Four-Year Colleges, by Field, Gender, and Academic Rank: 1995

Field of Doctorate/Sex	Total	Full Professor	Associate Professor	Assistant Professor	Instructor/ Teacher	Adjunct Faculty	Other Faculty	Does Not Apply
Total Sciences and Engineering (Number)	222,530	78,600	49,800	42,110	5,650	4,290	3,960	38,140
Male (Percent)	76.6	88.9	76.5	65.2	53.7	63.7	79.9	68.7
Female (Percent)	23.4	11.1	23.5	34.8	46.3	36.3	20.1	31.3
Total Sciences (Number)	196,870	69,280	43,380	37,380	5,290	3,820	3,430	34,120
Male (Percent)	74.4	87.6	73.8	62.6	51.0	60.7	77.1	66.3
Female (Percent)	25.6	12.4	26.2	37.4	49.0	39.3	22.9	33.7
Life and Related Sciences (Number)	72,120	22,780	14,780	14,710	1,920	1,240	1,510	15,180
Male (Percent)	70.3	85.5	68.5	60.2	56.4	62.2	69.1	61.5
Female (Percent)	29.7	14.5	31.5	39.8	43.6	37.8	30.9	38.5

Note: All numbers in table are estimates derived from a sample.

Source: "Doctoral Scientists and Engineers Employed in Universities and Four-Year Colleges, by Broad Field of Doctorate, Sex and Academic Rank: 1995," Table 17, National Science Foundation, Division of Science Resources Studies, *Characteristics of Doctoral Scientists and Engineers in the United States 1995*, NSF 97-319, R. Keith Wilkinson (Arlington, VA, 1997).

Among doctorate-holding scientists employed in the life sciences by universities and four-year colleges, more than half described their primary work activity as research and development, according to a 1995 National Science Foundation survey. Another 29 percent listed teaching as their primary activity. The remainder listed management, sales, and administration; computer applications; or other activities.

Salaries reflected the fact that those with higher levels of education were paid better than those with fewer academic credentials. Those 35- to 44-year-olds with doctorates or professional degrees (such as an M.D.) earned monthly incomes of $4,032–$6,537, compared with a monthly income of $2,891 for the same age group with only a bachelor's degree. On average, a 35- to 44-year-old who ended his or her formal education following a high school degree earned nearly $1,300 less per month than those holding a bachelor's degree, almost $1,600 less per month than master's degree recipients, and about $2,430 less in monthly income than did a doctoral degree recipient.

Average Monthly Income and Educational Attainment by Age, Spring 1993

	Monthly Income (in Dollars)					
	Total, 18 Years and Older		25–31 Years Old		35–44 Years Old	
	Standard		Standard		Standard	
Educational Attainment	Mean	Error	Mean	Error	Mean	Error
Doctorate	4,328	243.4	(B)	(B)	4,032	384.6
Professional	5,534	317.9	3,515	310.6	6,537	529.1
Master's Degree	3,411	122.9	2,648	150.9	3,180	124.1
Bachelor's Degree	2,625	52.2	2,341	54.1	2,891	97.6
Associate's Degree	1,985	64.2	1,760	74.1	2,145	103.2
Vocational Degree	1,736	63.5	1,643	105.6	2,030	148.9
Some College, No Degree	1,579	34.3	1,610	49.6	1,936	56.9
High School Graduate Only	1,380	18.7	1,310	26.4	1,603	49.9
Not a High School Graduate	906	17.9	936	54.1	1,032	49.6

Note
(B) Base less than 200 persons

Source: "What's It Worth? Field of Training and Economic Status: 1993," Current Population Reports, Census Bureau, December 1995.

The salaries of life sciences–trained bachelor's and master's degree recipients were also different in various careers in industry, academia, and government. Those whose educations ended with a bachelor's degree in biological sciences earned an average of $19,800 from private industry

Median Salary of Full-Time Employed 1994 Bachelor's Degree Recipients

| Major Field | Total | Broad Sector of Employment | | |
		Private Industry and Business	Educational Institution	Government
All Science and Engineering Fields	$24,000	$25,000	$20,000	$23,000
Life and Related Sciences, Total	$20,000	$20,000	$20,000	$21,000
Agricultural and Food Sciences	$20,000	$22,000	S	S
Biological Sciences	$19,800	$19,800	$20,000	$21,500
Environmental Life Sciences including Forestry Sciences	$20,000	$21,000	S	S

Note

S Data with weighted values less than 100 or unweighted sample sizes of less than 20 are suppressed for reasons of respondent confidentiality and/or data reliability.

Source: "Median Salary of Full-Time Employed 1994 Bachelor's Degree Recipients by Broad Sector of Employment and Field of Degree: April 1995," Chart B-55, National Science Foundation, Division of Science Resources Studies, *Characteristics of Recent Science and Engineering Graduates: 1995*, Detailed Statistical Tables, NSF 97-333, by John Tsapogas, Mary Collins, Lucinda Gray, and Debora Kraft (Arlington, VA, 1997), or see <http://www.nsf.gov/sbe/srs/nsf97333/start.htm>.

Median Salary of Full-Time Employed 1994 Master's Degree Recipients

| Major Field | Total | Broad Sector of Employment | | |
		Private Industry and Business	Educational Institution	Government
All Science and Engineering Fields	$38,000	$40,000	$30,000	$36,100
Life and Related Sciences, Total	$30,000	$33,000	$29,600	$28,000
Agricultural and Food Sciences	$30,000	$31,500	S	S
Biological Sciences	$30,000	$30,000	$29,600	S
Environmental Life Sciences including Forestry Sciences	$35,000	$36,000	S	S

Note

S Data with weighted values less than 100 or unweighted sample sizes of less than 20 are suppressed for reasons of respondent confidentiality and/or data reliability.

Source: "Median Salary of Full-Time Employed 1994 Master's Degree Recipients by Broad Sector of Employment and Field of Degree: April 1995," Chart B-111, National Science Foundation, Division of Science Resources Studies, *Characteristics of Recent Science and Engineering Graduates: 1995*, Detailed Statistical Tables, NSF 97-333, by John Tsapogas, Mary Collins, Lucinda Gray, and Debora Kraft (Arlington, VA, 1997), or see <http://www.nsf.gov/sbe/srs/nsf97333/start.htm>.

and business, educational institutions, and government, compared with an average of $30,000 for graduates holding master's degrees. Similar increases occurred between educational levels in other life sciences fields. Agricultural and food sciences and environmental life sciences averaged $20,000 for bachelor's degree recipients, but $30,000 and $35,000, respectively, for graduates with master's degrees.

RESEARCH AND DEVELOPMENT

Total research and development (R&D) expenditures at institutions of higher learning have shifted very little in terms of the percentage directed toward basic research and the amount for applied work, according to data from the National Science Foundation. In the two-decade span between fiscal years 1977 and 1997, the total amount of R&D expenditures increased from $2.8 billion to $16.7 billion, but the percentages directed toward basic and applied research changed very little.

When considering only federally financed R&D expenditures during that same two-decade period, the percentages directed toward basic and applied research remained about the same, but the total grew at a slower rate than R&D overall, which is funded by federal and other sources. Federally financed R&D rose from $2.0 billion to $10.5 billion during that period. In 1977, the federally financed portion of overall R&D expenditures made up more than 71 percent of the total, but in 1997, the federally financed portion amounted to about 63 percent of the total. The change in funding sources prompted many universities to seek research monies from non-federal sources, particularly industries, additional private ventures, and other nongovernmental associations and foundations.

An example of the shift in research funding came within the last decade when the federal government gave up its position as the greatest provider of health research funding, and industry took over that title. Another indicator of the shift was the difference in the number of young researchers whose grant applications were approved for funding from the federal government's granting giant, the National Institutes of Health. From 1985 to 1993, the percentage of young researchers (under 37 years old) who were able to win NIH grants dropped from 33 percent to 22 percent.

Research and Development R&D Expenditures at Universities and Colleges Over a Two-Decade Period

| | | Character of Work | | | |
| | | Basic Research | | Applied Research and Development | |
Fiscal Year	Total (in Millions)	Amount (in Millions)	Percent of Total	Amount (in Millions)	Percent of Total
1977	$4,067	$2,800	68.8	1,267	$31.2
1987	$12,153	$8,393	69.1	3,760	$30.9
1997	$24,348	$16,678	68.5	7,670	$31.5

Source: "R&D Expenditures at Universities and Colleges, by Character of Work: Fiscal Years 1953–97." Table B-2, National Science Foundation, Division of Science Resources Studies, Academic Research and Development Expenditures: Fiscal Year 1997, NSF 99-336, Project Officer, M. Marge Machen (Arlington, VA 1999), or see <http://www.nsf.gov/sbe/srs/nsf99336/start.htm>.

Federally Financed R&D Expenditures at Universities and Colleges Over a Two-Decade Period

| | | Character of Work | | | |
| | | Basic Research | | Applied Research and Development | |
Fiscal Year	Total (in Millions)	Amount (in Millions)	Percent of Total	Amount (in Millions)	Percent of Total
1977	$2,726	$2,007	73.6	$719	26.4
1987	$7,343	$5,375	73.2	$1,968	26.8
1997	$14,502	$10,484	72.3	$4,018	27.7

Note: R&D expenditures by character of work are based on estimates of basic research provided by universities and colleges.

Source: "R&D Expenditures at Universities and Colleges, by Science and Engineering Field: Fiscal Year 1997," Table B-39, National Science Foundation, Division of Science Resources Studies, Academic Research and Development Expenditures: Fiscal Year 1997, NSF 99-336, Project Officer, M. Marge Machen (Arlington, VA 1999), or see <http://www.nsf.gov/sbe/srs/nsf99336/start.htm>.

A snapshot of R&D expenditures at institutions of higher learning shows that from 1996 to 1997, the life sciences enjoyed a slightly greater increase, 6.7 percent, compared to the 5.4 percent rise in expenditures for all science and engineering fields. Within the life sciences, agricultural sciences jumped by only 3.3 percent, but biological sciences and medical sciences increased by 7.2 and 7.3 percent, respectively.

R&D Expenditures at Universities and Colleges, by Field and Fiscal Year (Dollars in Thousands)		
Field	1997 (in Thousands)	1996 (in Thousands)
All Science and Engineering Fields, Total	$24,348,336	$23,092,015
Engineering, Total	$3,818,493	$3,692,702
All Sciences, Total	$20,529,843	$19,399,313
Life Sciences, Total	$13,607,902	$12,756,030
Agricultural Sciences	$1,979,467	$1,916,169
Biological Sciences	$4,227,343	$3,941,202
Medical Sciences	$6,866,902	$6,396,351
Other	$534,187	$502,308

Source: "R&D Expenditures at Universities and Colleges, by Source of Funds and Science and Engineering Field: Fiscal Years 1990–97," Table B-3, National Science Foundation, Division of Science Resources Studies, Academic Research and Development Expenditures: Fiscal Year 1997, NSF 99-336, Project Officer, M. Marge Machen (Arlington, VA 1999), or see <http://www.nsf.gov/sbe/srs/nsf99336/start.htm>.

Reference

Beardsley, T., "Big-Time Biology," *Scientific American*, November 1994: 90–97.

CHAPTER ELEVEN
Resources

B iologists have many resources available to them. The organizations and associations listed in Chapter 12 are excellent sources of information in their journals and other publications alone. For more detailed information about particular topics and current research, most of the groups hold annual meetings and run workshops and symposia. A number of the organizations and associations have subgroups or committees that have further specializations.

In addition, many also have online discussion groups, which allow biologists and other scientists in specialty areas to converse publicly by e-mail. Anyone who joins the discussion group can then either participate in the discussion or simply read the exchanges. Often, a properly targeted, online discussion group is the best place to learn about new research, publications, controversies, policy changes, and other specific topics. A good starting point to find the discussion group most appropriate to a particular line of questioning is through the Web sites of related organizations. If the Web site doesn't refer to a specific discussion group, a quick note to the e-mail address provided on the site—or to the site's webmaster—will usually result in a few recommendations.

Besides these member organizations and associations, publications, and discussion groups, biologists will find a bounty of resources at the library, in bookstores, and on the World Wide Web. This chapter provides a listing of some of what is available and also describes the services offered by a few other organizations that have not been included elsewhere in this book.

ORGANIZATIONS

In addition to the member organizations and associations mentioned in Chapter 12, several government agencies and organizations can be great sources of information for biologists. This section will include information about a few of those agencies and organizations. Web addresses (URLs) occasionally change, so if a URL provided below is outdated, use a search engine to locate the new one.

Environmental Protection Agency (EPA)

The mission of the Environmental Protection Agency (EPA)—to protect human health and the environment—has obvious overlaps with the interests of many biologists.

Both the EPA's main World Wide Web site at <http://www.epa.gov/>, and its site for scientists and researchers at <http://www.epa.gov/epahome/research.htm>, can direct biologists to information such as environmental data and tools for scientific inquiry, the agency's ongoing scientific and research-related programs, research grants and fellowships, possibilities for research collaboration, and EPA reports and other publications. The science site also directs biologists to the EPA's Center for Environmental Information and Statistics, to the access point for data from five major EPA databases, and to the flora- and fauna-based Integrated Taxonomic Information System.

National Aeronautics and Space Administration (NASA)

The National Aeronautics and Space Administration may have its most obvious tie with astronomers, but biologists can also find its resources helpful. One of NASA's offerings that is of special interest to biologists is its Mission to Planet Earth. As part of that mission, NASA launched the satellite SeaWiFS to monitor changes in phytoplankton, the tiny plants that drift in the surface layer of a sea or lake. Phytoplankton absorb atmospheric carbon dioxide, which is one of the greenhouses gases produced by the burning of fossil fuels, and therefore may provide clues about climate change. Those who wish to view images from the satellite should refer to the Web site maintained by NASA at <http://seawifs.gsfc.nasa.gov/SEAWIFS/IMAGES/SEAWIFS_GALLERY.html>.

The main NASA Web site maintains a multimedia gallery of images, located specifically at <http://www.nasa.gov/gallery/index.html/>. Astrobiologists, who study the possibility of extraterrestrial life, are especially interested in Mars and in Jupiter's moon Europa. They believe the evidence of water on those worlds makes them potential spawning grounds for life. NASA maintains a Web site, <http://astrobiology.arc.nasa.gov/index.cfm>, that includes a description of this young science, information on upcoming conferences and workshops, an online astrobiology forum, related articles, and links to additional resources.

In addition, biologists and biology students will find data and images on NASA's Earth Observatory Web site at <http://earthobservatory.nasa.gov/>. This site contains data on such global and regional changes as sea temperatures, vegetation cover, and weather patterns. It also includes articles and explanations of various phenomena, projects animations that clarify the topics under discussion, and presents opportunities to consider the effects of environmental changes.

National Library for the Environment
In July 1999, the National Library for the Environment launched the Environmental Research Information Exchange (ERIE), a forum for environmental scientists, students, policymakers, and others. Located on the Web site for the Committee for the National Institute for the Environment, the forum at <http://www.cnie.org/exchange.htm> provides a bulletin board and highlights page for posting information about research information and needs, job openings, funding opportunities, and collaboration possibilities. Both researchers and those who seek input from researchers will find ERIE a helpful resource.

National Library of Medicine (NLM)
Part of the National Institutes of Health, the National Library of Medicine has a slew of resources available to biologists. Particularly good starting points for biologists are the library's Visible Human Project at <http://www.nlm.nih.gov/research/visible/visible_human.html> and the Genes and Disease Web site at <http://www.ncbi.nlm.nih.gov/disease/>.

The Visible Human Project will eventually provide complete, anatomically detailed, three-dimensional, virtual human bodies. Until that proposed goal is realized, the site offers a wide range of data and links to other visible human projects. The Genes and Disease Web site provides nontechnical information about the relationship between genes and disease. The Genes and Disease site explains how genes can alter biological pathways and lead to several dozen diseases, and provides links to additional detail, such as a view of these genes' locations on chromosomes.

National Oceanic and Atmospheric Administration (NOAA)

The Web site of the National Oceanic and Atmospheric Administration (NOAA) at <http://www.noaa.gov/> provides links to each of its major organizations: the National Weather Service, the National Environmental Satellite, the Data and Information Service, the National Marine Fisheries Service, the National Ocean Service, and the Office of Oceanic and Atmospheric Research. NOAA also maintains a large photographic archive containing images of islands, coastal regions, deep-sea life, and coral reef habitats.

National Science Foundation/Long-Term Ecological Research (LTER)

The National Science Foundation sponsors 21 Long-Term Ecological Research (LTER) sites with the purpose of "investigating ecological processes operating at long time scales and over broad spatial scales." More than 1,100 scientists and students take part in the LTER network <http://lternet.edu/>.

For example, the Baltimore Ecosystem Study is one of two LTERs with an urban area as a focal point. The Baltimore study began in 1997 with nearly three dozen scientists who began to audit the metropolitan area and Chesapeake Bay as an ecological system with various forces acting upon it, such as urbanization, urban revitalization, and invasion by non-native plant species.

According to Steward Pickett, an ecologist at the Institute of Ecosystem Studies in Millbrook, New York, and principal investigator and project director of the Baltimore study, the area is diverse ecologically, holds numerous watersheds, and contains a significant land-sea interface. By studying an area over a lengthier period of time, scientists hope to gain critical information about the effects of urbanization on socioeconomic and ecological changes.

New York Botanical Garden

For a view of botanical resources, the botanical community often turns to the New York Botanical Garden's Index to American Botanical Literature. The index, which was originated in 1886 by the editor of the *Bulletin of the Torrey Botanical Club*, provides information on books and articles dealing with botany. Since 1999, the index has become an electronic resource, available at <http://www.nybg.org/bsci/iabl.html>. The index currently begins with articles published in 1996, although efforts are under way to go further back in history.

The index includes articles dealing with plants from Greenland to Antarctica, and topics ranging from such areas as fossil and current plants to systematics, ecology, and economic botany.

North American Reporting Center for Amphibian Malformations (NARCAM)

The North American Reporting Center for Amphibian Malformations (NARCAM) is a clearinghouse for information about deformities among frogs and salamanders. The center's Web site at <http://www. npwrc.usgs.gov/narcam> provides information by county about the incidence of amphibian malformations. The site also contains photographs of malformations, including abnormal, extra, or missing limbs; missing eyes; and radiographs of malformed individuals. The NARCAM site has online forms for both scientists and the public to submit sightings about malformed amphibians to add to the database of information.

By making this information available, NARCAM hopes to encourage understanding of the problem, promote research into these reported deformities, and provide a catalyst for possible research collaborations.

U.S. Geological Survey (USGS) Biological Resources Division (BRD)

The National Biological Service is now the Biological Resources Division of the U.S. Geological Survey. This latest switch, which occurred in 1996, is one in a long line of changes for the agency.

The Biological Resources Division traces its roots to 1885 and the Ornithological Office within the Entomology Division of the U.S. Department of Agriculture. A year later, it expanded to the Division of Ornithology and Mammalogy, and in 1896 it became the Division of Biological Survey. In less than a decade, the division was converted to a bureau. In 1939, the Bureau of Biological Survey was transferred to the U.S. Fish and Wildlife Service at the Department of the Interior, where it remained until 1993.

In 1993, the unit transformed into the National Biological Service, and three years later became part of the U.S. Geological Survey as the Biological Resources Division (BRD). Its mission is "to work with others to provide the scientific understanding and technologies needed to support the sound management and conservation of our nation's biological resources."

One portion of the new BRD that biologists are finding particularly useful is the National Biological Information Infrastructure (see Chapter 3), which brings together biology-related databases and informational resources. That information is not only available to government agencies, but to researchers, students, and others with an interest in biology.

In addition, the BRD has a variety of national programs, including the Arctic Science Program, the Bird Banding Laboratory, the Biomonitoring Environmental Status and Trends program, the Global Change Research Program, and the three-year, teacher-oriented pilot program called Earth Stewards. Its Integrated Taxonomic Information System provides a standard reference for plant and animal taxonomy and nomenclature, while the North American Breeding Bird Survey shares information about population trends and distribution. These are just a few of the many programs and information resources available through the Biological Resources Division.

Information about these programs and the BRD are available at the division's web site, <http://www.nbs.gov/>, or by writing to the Biological Resources Division–USGS, U.S. Department of the Interior, Office of Public Affairs, 12201 Sunrise Valley Dr., Reston, VA 20192.

WEB SITES

This section lists a sample of the many Web sites run by universities and major organizations. While limited in scope, this section should give readers a sense of the variety available on the World Wide Web. The Web sites listed below, along with the Web sites of numerous organizations and associations (see Chapter 12), may also provide links to additional sites of interest.

BioTech
http://biotech.icmb.utexas.edu/
Created by an evolutionary engineer, the BioTech Web site is most notably an index to Web resources relating to biology, chemistry, and biotechnology. In addition, the site directs users to a variety of other resources, including a large life-sciences dictionary and a botanical resource known as Cyberbotanica that spells out the medicinal uses for plants.

Dino Russ's Lair
http://www.isgs.uiuc.edu/isgsroot/dinos/dinos_home.html
Dino Russ's Lair, a site managed by a member of the Illinois State Geological Survey, provides an index to the ever-increasing number of Web sites highlighting dinosaurs. Links to research articles and techniques, art featuring dinosaurs and other vertebrates, museum exhibits, and online information about paleontological digs are included.

DNA Vaccine Web
http://www.genweb.com/Dnavax/dnavax.html
The DNA Vaccine Web provides the latest on the development of vaccines against disease. Created and updated by a vaccinologist, the site provides a list of hundreds of related papers, along with a news archives, a meeting calendar, and updates on research techniques and other topics.

Image Library of Biological Macromolecules
http://www.imb-jena.de/IMAGE.html
With thousands of images of proteins and other molecules, the Image Library of Biological Macromolecules is a resource for scientists and nonprofessionals alike. Hosted by the Institute of Molecular Biotechnology in Germany, the library's visual offerings include images, drawings, and various types of models. The site also provides the basics about structural biology and associated research techniques.

Liszt
http://www.liszt.com
Liszt provides a directory of electronic mailing lists. As of June 28, 1999, Liszt provided descriptions of 90,095 mailing lists, along with instructions on subscribing to each of them. In addition to the directory, which is arranged by broad topics, Liszt offers a search service to find lists of interest to the user.

Biologists might find the nature and science topics useful. Topics of the mailing lists under the biology grouping, for example, include lists specializing in such areas as bees, ancient DNA, plant cell wall biology, genetic algorithms and neural networks, orchids, and postdoctoral positions.

Online Macromolecular Museum
http://www.clunet.edu/BioDev/omm/gallery.htm
The Online Macromolecular Museum not only provides explanations of the components of molecules, but accompanies them with images of the molecules, color highlights to indicate the precise location of the component under consideration, the ability to zoom in and to rotate the molecules, and links to additional information.

Paleontology without Walls
http://www.ucmp.berkeley.edu/
This popular site, created through the University of California–Berkeley's Museum of Paleontology, features sections on animals and plants, time periods, phylogeny, geology, and evolution, in addition to special and new exhibits. Of the more than 1,000 exhibits within the site, many contain photographs and some offer short movies to accompany the descriptions.

The Entomology Index of Internet Resources
http://www.ent.iastate.edu/list/
Hosted by Iowa State University, the Entomology Index of Internet Resources arranges more than 1,000 links to insect-related Web sites into an all-text index. Sections on such topics as beekeeping, insect sounds,

medical entomology, and integrated pest management help visitors find their way through the site.

The Tree of Life
http://phylogeny.arizona.edu/tree/phylogeny.html
With input from several hundred scientists, University of Arizona systematists are trying to generate an expansive phylogenetic tree that contains all of the Earth's known organisms. As of mid-July 1999, the pages in the Tree of Life project were housed on 20 computers in four countries.

Some of the branches on the tree are fuller than others, as scientists have added information about organisms within their specialties. In some cases, the branches fan out to individual species and incorporate photographs and elaborate descriptions.

Universal Viral Database
http://life.anu.edu.au/viruses/welcome.htm
Provided by the International Committee on Taxonomy of Viruses, the Universal Viral Database site provides information about thousands of viruses. The site also describes the life cycles of many of the viruses featured. Diagrams and images of some of the viruses are also included.

Vascular Plant Image Gallery
http://www.csdl.tamu.edu/FLORA/gallery.htm
Botanists at Texas A&M University maintain the Vascular Plant Image Gallery, a compilation of thousands of plant photographs arranged by family. Besides providing a view of some of nature's most intricate creations, the site gives botany students and taxonomists an additional resource for educational or plant-identification purposes.

BOOKS

While the World Wide Web, journals, and magazines can help keep readers up to date on current research, books can take that information; collate it; summarize it; compare it; provide insights about its significance, context, and ramifications; and help the reader to truly understand the subject at hand. As many biologists become more specialized in their work, books can also help them stay in touch with the field overall, and perhaps even provide new avenues and previously unseen connections for their own research.

This section lists titles and short descriptions of a variety of current biology-related books of particular value.

About This Life: Journeys on the Threshold of Memory
by Barry Lopez, Alfred A. Knopf, 1998

Listed as one of the top 10 nature books on the National Audubon Society's Web site <http://www.audubon.org/market/publish/top10.html>, *About This Life: Journeys on the Threshold of Memory* is a collection of autobiographical essays that portray the ideas and adventures of award-winning author Barry Lopez. The essays take the reader from Alaska to Antarctica and from the Galapagos to Japan. The *Publisher's Weekly* review reports, "For Lopez, the world's topography is memory made manifest; it stimulates Lopez's own recall and that, in turn, forces us to really think." *Kirkus Reviews* adds, "Lopez ventures forth, hunts and gathers the sacred twinings of humanity and nature, and returns with stories as venerable as the best folktales."

Biomimicry: Innovation Inspired by Nature
by Janine M. Benyus, William Morrow & Co., 1997

Biomimicry: Innovation Inspired by Nature introduces readers to the scientists and the science behind the race to mimic biological processes for uses as wide ranging as agriculture, pharmaceuticals, and technology. Author Janine M. Benyus follows scientists who question whether nature holds the key to pesticide-free crops, human disease cures, and more efficient energy systems. A *Booklist* review noted, "For Benyus, though, a technology that mirrors nature does more than enlarge human powers and gratify human ambitions. Such a technology teaches us how to live in harmony with nature, rather than how to dominate it."

Biotechnology Unzipped: Promises and Realities
by Eric S. Grace, National Academy Press, 1997

After an introduction to genetics and genetic engineering, author Eric S. Grace delves into the application of and controversies surrounding biotechnology, including cloning and DNA research. He reminds the reader that biotechnology is a commercial enterprise, rather than a philanthropic exercise. A review in the *New York Times Book Review* stated, "Since 'the focus of biotechnology companies is profit, not philanthropy,' *Biotechnology Unzipped* serves to caution us that without the requisite judiciousness, even the most wondrous and potentially life-saving feats of science 'cannot, in the end, save us from who we are.'"

Broadsides from the Other Orders: A Book of Bugs
by Sue Hubbell, Houghton Mifflin Co., 1998

"The book contains more than enough technical information to be useful to biology students, while at the same time will hold the interest of the merely curious," states a review in *School Library Journal*. *Broadsides from the Other Orders: A Book of Bugs* describes 13 common insects, including gypsy moths and killer bees, and includes information about evolution, taxonomy, physiology, and other aspects of these inverte-brates, while debunking the myths that surround them. The National Audubon Society's Web site lists *Broadsides from the Other Orders* as one of the top 10 nature books.

Charles Darwin's Letters: A Selection, 1825–1859
Frederick Burkhardt (ed.), Cambridge University Press, 1996

This collection of letters written by Charles Darwin provides a different view of the famed scientist who with fellow naturalist Alfred Wallace formulated modern evolutionary theory. The collection includes both reflective and scientific letters before, during, and after his historic voyage on HMS *Beagle*, where he made crucial observations that contrib-uted to his ideas about natural selection and evolution. The special 1998 edition of *Science Books and Films* reviews *Charles Darwin's Letters* with this comment: "In the age of e-mail, when CD-ROM collections of ready-made correspondence have removed the need for creative thought from the art of writing letters, it is refreshing that the handwritten letter can still be so compelling."

Civilization and the Limpet
by Martin J. Wells, Perseus Books, 1998

Zoologist and Cambridge professor Martin J. Wells describes the odd, unusual, and interesting organisms of the sea in this book of essays. Using marine animals for illustration, Wells paints a picture of the diversity of life in the oceans, touching upon such topics as reproduction, navigation, and evolution in his book *Civilization and the Limpet*. Recommended on the National Audubon Society's Web site, the book has also received praise from numerous reviewers. Said *Library Journal*, "Begun as separate articles while Wells was sailing from Southampton, England, to France, the chapters intersperse personal anecdotes and accounts of Wells's scientific background and experiences with the facts of marine biology." *Publisher's Weekly* added, "He defends a career in biology as one that is never boring—a claim borne out by his winsome book."

Darwin's Dangerous Idea: Evolution and the Meanings of Life
by Daniel Clement Dennett, Touchstone Books, 1996

Of Daniel Clement Dennett's *Darwin's Dangerous Idea: Evolution and the Meanings of Life*, a *Wall Street Journal* review noted, "Dennett is a philosopher of rare originality, rigor and wit. Here he does one of the things philosophers are supposed to be good at: clearing up conceptual muddles in the sciences." *Darwin's Dangerous Idea* presents the intrica-cies of evolution and natural selection, as well as some of the controver-sies, and the importance and implications of evolutionary thought. The book was a 1996 Pulitzer Prize finalist for general nonfiction.

Dinosaur in a Haystack: Reflections in Natural History
by Stephen Jay Gould, Harmony Books, 1996

One of the best-known biologists of this era, Stephen Jay Gould presents nearly three dozen essays about such diverse topics as the beginning of the millennium, the influence of Hollywood on the public's perception of evolution, and the paleontology of snails. The essays were previously published in his weekly *Natural History* magazine column. Of *Dinosaur in a Haystack: Reflections in Natural History*, the *New York Times Book Review* remarked, "He is an incomparable explainer of difficult ideas."

Earth Odyssey: Around the World in Search of Our Environmental Future
by Mark Hertsgaard, Random House, 1998

In *Earth Odyssey: Around the World in Search of Our Environmental Future*, author and journalist Mark Hertsgaard reports on his findings during a six-year, 19-country journey around the world. Along the way, he attempts to determine whether the future of the human race is at risk. A review in *Booknews* states, "He reports on the global environmental predicament through the eyes of people who live with it, through interviews with politicians and peasants, taxi drivers and activists." The National Audubon Society's Web site recommends the volume as one of the top 10 nature books.

Guns, Germs and Steel: The Fates of Human Societies
by Jared Diamond, W. W. Norton & Co., 1997

Winner of the 1998 Pulitzer Prize for general nonfiction, *Guns, Germs and Steel: The Fates of Human Societies* peers into the reasons human societies are the way they are. Author Jared Diamond, a physiologist at UCLA, explains how geography, demography, and ecology have all influenced global human history. Reviews have described *Guns, Germs and Steel:* as "evenhanded," "astounding," and "highly important."

How the Mind Works
by Steven Pinker, Norton, 1997

Psycholinguist Steven Pinker's *How the Mind Works* was a finalist for the 1998 Pulitzer Prize for general nonfiction and a finalist for a 1997 National Book Critics Circle Award for general nonfiction. Combining views of cognitive scientists and sociobiologists, this book offers insight into the intricacies of the mind, and introduces the role of evolution to the mind's cognitive function. *The New York Times Book Review* stated, "Pinker has breathed marvelous life into the computational models, the originals of which are buried in nerdish obscurity."

How We Die: Reflections on Life's Final Chapter
by Sherwin B. Nuland, Knopf, 1994

A Pulitzer Prize finalist for general nonfiction, *How We Die: Reflections on Life's Final Chapter* characterizes six common fatal conditions, then delves into how the different players involved—patients, health professionals and administrators, and family members—deal with impending death. Written by physician Sherwin B. Nuland, the book details the physical processes that accompany a body's demise, and removes the myths from this often-sugar-coated subject. A reviewer in *Kirkus Reviews* commented, "Nuland selects several common causes of death—heart attack, old age, Alzheimer's, violence, AIDS, and cancer—and, with unrelenting honesty and unsettling detail, shows precisely what happens to the body involved."

Huxley: From Devil's Disciple to Evolution's High Priest
by Adrian Desmond, Addison-Wesley, 1997

Scientist and physician Thomas Henry Huxley took a wild ride in the mid- to late-1800s, embracing evolution while bringing science into the homes of ordinary citizens. *Huxley: From Devil's Disciple to Evolution's High Priest* documents that ride and provides insight into an era when society struggled with—and often passionately fought against—notions that humans descended from apes. According to the review in the 1998 special edition of *Science Books and Films*, "The volume gives entertaining and informative facts and sidelights, not only into Huxley's personal and scientific life, but also into the life of his family and of Darwin." (Charles Darwin and Alfred Wallace were naturalists who independently formulated the basis for the modern theory of evolution.)

Life: A Natural History of the First Four Billion Years of Life on Earth
by Richard Fortey, Knopf, 1998

Taking his cues from the fossil evidence, paleontologist Richard Fortey travels up the evolutionary tree. While describing evolution, he takes the reader on numerous field adventures with scientists and field assistants. Politics and academic ambition also have parts in this book. *Kirkus Reviews* noted, "His wonderful description of the emergence and proliferation of life on Earth combines the vision of a scientist with an intimate knowledge of the fossil record with the insight of a scholar for masterful interpretation."

Naturalist
by E. O. Wilson, Island Press, 1994

An autobiography, *Naturalist* invites the reader into the life of one of today's best-known evolutionary biologists and naturalists. E. O. Wilson describes his fascination with nature as a string that has run throughout his life from his childhood into his days as an academician at Harvard University. Wilson also explains the impact that ants and other social insects have had on his scientific career. *Science Books and Films* describes *Naturalist* as "stimulating" and "enlightening," and also recommends it as "a good intellectual background" for other books by Wilson.

Our Molecular Nature: The Body's Motors, Machines and Messages by David S. Goodsell, Copernicus Books, 1996

Filled with illustrations, *Our Molecular Nature: The Body's Motors, Machines and Messages* takes readers on a tour of the molecules essential to life. Scientist and author David S. Goodsell provides descriptions of such molecules as enzymes, toxins, and hormones, along with insights into why some molecules can have such drastic impacts on living things. He also gives readers a sense of how quickly humans' knowledge of molecules has grown in recent years.

Picture Control: The Electron Microscope and the Transformation of Biology in America, 1940–1960 by Nicolas Rasmussen, Stanford University Press, 1997

Picture Control: The Electron Microscope and the Transformation of Biology in America, 1940–1960 takes readers through the explosion in biological sciences that occurred in the middle of the twentieth century. The biological sciences saw whole new fields develop during this two-decade period of technological and political upheaval around the globe. A review in *Science* summarized the book's lessons as "how the threads of the past influence or hinder the ways by which we learn in science to see with a fresh eye."

Rattlesnake: Portrait of a Predator by Manny Rubio, Smithsonian Institution Press, 1998

Filled with color photographs, *Rattlesnake: Portrait of a Predator* offers an in-depth view of this group of intriguing and often-feared animals. Professional and amateur herpetologists will find information about evolution, physiology, and ecology, as well as a listing of herpetological societies. A review in the January/February issue of *Science Books & Film* states, "Snake lovers will enjoy this book. Those willing to put out a little effort will learn much from the text; others may find it sufficient to just enjoy the illustrations."

Retroviruses edited by John M. Coffin, Stephen H. Hughes, and Harold E. Varmus, Cold Spring Harbor Laboratory Press, 1998

Described in one review as the new classic in the field, *Retroviruses* contains chapters from nearly two dozen researchers, plus an assortment of illustrations and diagrams. The book includes a wide range of topics spanning the origins and classifications of retroviruses, structure and function, pathogenesis, and other aspects of retroviruses. A review in *Science* commented, "*Retroviruses* conjures up that bookshop feeling—the exhaustion and pleasure after having spent the best part of an afternoon squatting on the floor, absorbed in the book in front of you."

Skeptics and True Believers: The Exhilarating Connection Between Science and Religion
by Chet Raymo, Walker, 1998

"*Skeptics and True Believers* raises profound questions about how we face, filter and deny realities about the universe as received from science," stated a review in *Science*. Author Chet Raymo considers the two opposing views of the "forces" governing life in *Skeptics and True Believers: The Exhilarating Connection Between Science and Religion*. The book takes into account the history of science, as well as the interplay of new knowledge vs. new mysteries, in exploring the age-old struggle between beliefs and proven facts.

Song for the Blue Ocean
by Carl Safina, Henry Holt & Co., 1997

Author Carl Safina, scientist and founder of National Audubon Society's Living Oceans Program, describes how the oceans are tied to human survival in *Song for the Blue Ocean*. From the coasts to the ocean's greatest depths, he examines marine habitats the world over and calls attention to the global declines in fish populations. The *New York Times Book Review* provided this commentary: "an engrossing, illuminating, depressing journey, with a research ecologist, to the dwindling populations of wild edible creatures in the sea." The National Audubon Society's Web site recognizes *Song for the Blue Ocean* as one of its top 10 nature books.

The Beak of the Finch: A Story of Evolution in Our Time
by Jonathan Weiner, Alfred A. Knopf, 1994

In *The Beak of the Finch: A Story of Evolution in Our Time*, the author provides an in-depth introduction to, and review of, evolution, mainly by chronicling the work of two modern-day biologists—Peter and Rosemary Grant—on Daphne Major in the Galapagos Islands. The Grants spent years banding and then following subtle shifts in the species of the island's finches over time and through environmental changes. *Booklist* describes the book as "an engaging account of a seminal study that introduces the reader to Darwin and to the dedicated, tireless biologists who have proved him right." *The Beak of the Finch* won the 1995 Pulitzer Prize for general nonfiction.

The Coming Plague: Newly Emerging Diseases in a World Out of Balance
by Laurie Garrett, Farrar Straus & Giroux, 1994
Five decades of war between humans and microbes are chronicled in *The Coming Plague: Newly Emerging Diseases in a World Out of Balance*, which was recognized as a *New York Times* Notable Book in 1994. *The Coming Plague* combines scientific literature and numerous interviews with accounts of the spread of various diseases, including AIDS, the ebola and hanta viruses, and legionnaire's disease. In this book, National Association of Science Writers member Laurie Garrett presents a review of the evolution of microbes, and suggests that the human battle is far from over. *Kirkus Reviews* remarked, "One does not like to apply the phrase too often in a book review, but here is a volume that should be required reading for policymakers and health professionals."

The Diversity of Life
by E. O. Wilson, Norton, 1993
Pulitzer Prize–winning author E. O. Wilson describes a myriad of biological concepts in this book about evolution and species diversity. *The Diversity of Life* covers such topics as food webs, speciation, the number of species on Earth, habitat types, and medicinal plants, while making broader points and drawing such conclusions as: "Every country has three forms of wealth: material, cultural and biological. … The essence of the biodiversity problem is that biological wealth is taken much less seriously. This is a major strategic error, one that will be increasingly regretted as time passes." Reviews describe the book as "important," "stirring" and "deftly written."

The Double Helix: A Personal Account of the Discovery of the Structure of DNA
by James D. Watson and Gunther S. Stent, Simon and Schuster, 1998
James D. Watson was one of three scientists to share a Nobel Prize for their work in unraveling the structure of the DNA molecule. *The Double Helix: A Personal Account of the Discovery of the Structure of DNA* provides Watson's view of the inside story behind their work. The New York Review of Books' *Readers Catalog* describes the book as "a personal, candid memoir of the discovery of the structure of DNA."

The Life of Birds
by David Attenborough, Princeton University Press, 1998
Listed as one of the National Audubon Society's picks on its Web site for the top 10 nature books, *The Life of Birds* is designed to accompany the 1999 PBS television series of the same name. The series and the book portray the behavior, adaptation and evolution, and many other aspects

of these winged creatures in what one review described as lively and straightforward prose from well-known writer David Attenborough.

The Work of Nature: How the Diversity of Life Sustains Us
by Yvonne Baskin, Island Press, 1997
A review in the journal *Nature* defined *The Work of Nature: How the Diversity of Life Sustains Us* as a "must read" for citizens at all levels of expertise from students to teachers and scientists. This book, by National Association of Science Writers member Yvonne Baskin, takes a look at biodiversity and its importance to the health and function of the Earth's ecosystem.

Volvox: Molecular-Genetic Origins of Multicellularity and Cellular Differentiation
by David L. Kirk, Cambridge University Press, 1998
The green algae-like *Volvox* has become a model system for multicellularity and provided biologists with insights into development. In *Volvox: Molecular-Genetic Origins of Multicellularity and Cellular Differentiation*, David L. Kirk presents findings from his lab and similarly inclined research teams, as well as evolutionary and ecological analyses. A review in *Science* proclaimed, "Kirk ... gives authoritative accounts of both the molecular genetics of *Volvox* development (where his own contributions have been made) and the evolutionary genetics and ecology of the volvocine algae. ... [The book is] a synthesis that can be appreciated and enjoyed by any biologist."

OTHER ELECTRONIC RESOURCES

CD-ROMs may accompany books or may stand alone as biology resources. This section takes a look at a few of the CD-ROM resources that might be of interest to biology professionals and students.

Magazines and Journals

Some science and biology magazines and journals may provide back issues on compact disc. For example, *Science News*, a weekly news magazine, offers the previous four years of its issues on CD-ROM. The disk contains the magazine's complete text, including news stories, features, notes, and letters to the editor. All of the material is searchable. Information about this product is available through the *Science News* Web site at <http://www.sciencenews.org/> or by calling 800-544-4565.

Most magazines and journals list whether they offer this option on their Web sites. The organizations and associations listed in the next chapter provide the names of the journals and newsletters associated with them, along with URLs for their main Web sites.

Databases

Databases of biological literature are available on CD from a variety of sources, but the more extensive databases carry high price tags (generally several thousand dollars or more).

Most science libraries have a number of these databases available for use by the public. Biological Abstracts through BIOSYS, for example, provides more than 5.7 million life-sciences records from 5,500 international journals dating from 1980. It covers everything from anatomy to botany and ecology to genetics. Biological Abstracts/RRM contains nearly 160,000 references to meetings, symposia, and workshops, plus another 20,000 references to review articles.

Other databases are more specialized. For instance, Elsevier Science Ltd.'s ECODISC offers some 85,000 ecological abstracts and records from more than 2,000 journals, books, and other sources. Another example is PARASITECD, which covers parasitology. This database, produced by CABI Publishing, includes material dating back to 1973.

Other Compact Disks

Depending on specific interests and level of specialty, biologists and students may find many or few compact discs available to meet their needs.

Generally, nature-related CDs are numerous, particularly those directed at children. Examples of nature-related disks are *The Butterflies of North America* CD-ROM, by James A. Scott, and *The National Audubon Society Interactive CD-ROM Guide to North American Birds*. Both received praise from reviewers. The butterfly CD includes videos, music, and additional photos and text, along with information about butterfly physiology and behavior, and tips on raising and collecting them. The bird CD provides videos, photos, songs and calls, plus maps and trip planners.

Biologists and students should be able to receive recommendations for pertinent and useful compact discs from colleagues either at their home institutions, or perhaps via electronic discussion groups. Computer magazines also often print reviews of various products, and might be helpful when selecting a CD to fit a particular need.

CHAPTER TWELVE
Organizations and Associations

T he number of organizations and associations that deal with biology ranges into the hundreds. The listing in this chapter is by no means complete, but should provide an indication of the immense diversity of groups available to biology students, teachers and professors, researchers, field biologists, and lab technicians. The chapter begins with some of the more broad-based groups, then discusses more specialized organizations and associations.

BROAD-BASED GROUPS

American Association for the Advancement of Science (AAAS)
1200 New York Avenue, NW
Washington, DC 20005
telephone: 202-326-6400
e-mail: webmaster@aaas.org
URL: http://www.aaas.org/

The American Association for the Advancement of Science (AAAS) formed in 1848 to promote all sciences and engineering. Since then, the association's annual weeklong meetings have expanded, now drawing thousands of scientists, engineers, and students from the United States and abroad to present papers (and posters, in the case of the students), or to listen to symposia and lectures featuring some of the most respected scientists in the world.

Directors of many of the nation's top science organizations, including the National Aeronautics and Space Administration (NASA), the National Science Foundation (NSF), and others, routinely attend the meetings. In addition, numerous well-known speakers, such as writer Michael Crichton, computer guru Bill Gates of Microsoft, President Bill Clinton, and Vice President Al Gore, have recently presented talks on various topics. The meetings also attract hundreds of reporters who cover the sessions for the national and international media.

Besides its annual meetings, AAAS publishes the widely known and respected weekly journal *Science*, which includes a news section, research articles, book reviews, and an extensive section of classified advertisements for jobs in the public and private sectors. The association helped kick off the television program *Nova* and now offers the science-adventure radio program "Kinetic City Super Crew" for children aged 7 to 12.

AAAS also is involved in many science-related pursuits, such as encouraging members of underrepresented groups to consider science as a career, improving the public's knowledge about science, becoming involved in national policy decisions, promoting international scientific exchange, and heightening both the media's understanding of science and scientists, and scientists' understanding of the media. In its Project 2061, the AAAS is trying to reform math and science education from kindergarten through high school so that all Americans become science-literate. The Project 2061 had its start in 1985, the last appearance of Halley's Comet; 2061 is the year the comet will return.

American Institute of Biological Sciences (AIBS)
1444 I Street, NW, Suite 200
Washington, DC 20005
telephone: 202-628-1500 or 800-992-2427
fax: 202-628-1509
URL: http://www.aibs.org/

More than five decades old, the 6,000-member American Institute of Biological Sciences is a melting pot for the nation's biologists. If the members of its 50 member societies are added to the institute's membership, the total increases to 125,000. The organization is known for both its peer-reviewed magazine, *BioScience*, and its extensive and comprehensive meetings, particularly its cross-disciplinary annual meeting that injects basic and applied science into its sessions. Members of the AIBS can take advantage of outreach programs and publishers' discounts.

American Society for Bioethics and Humanities (ASBH)

4700 W. Lake
Glenview, IL 60025-1485
telephone: 847-375-4745
fax: 847-375-6345
e-mail: info@asbh.org
URL: http://www.asbh.org/

This professional society invites anyone engaged in work that touches on clinical and academic bioethics or the health-related humanities to join its membership. Established in 1998, the American Society for Bioethics and Humanities was the result of a consolidation of three groups: the American Association of Bioethics, the Society for Bioethics Consultation, and the Society for Health and Human Values. The society promotes education and research in the areas of bioethics and humanities and fosters discussions in policy-making arenas. Its annual meeting, like its membership, is multidisciplinary. Members of the ASBH receive the society's quarterly publication, *ASBH Exchange,* which provides various articles, news updates, and event listings. In addition, the society has a number of affinity groups, or special-interest groups, that participate in educational and research issues and in legislative policy-making.

Association for Biology Laboratory Education (ABLE)

Nancy Rosenbaum (ABLE Membership Chair)
Department of Biology
Yale University
P.O. Box 208104
New Haven, CT 06520-8104
telephone: 203-432-3864
e-mail: nancy.rosenbaum@yale.edu
URL: http://www.zoo.utoronto.ca/able/

Founded in 1979, the Association for Biology Laboratory Education encourages university and college educators to share experiments and exercises that will improve the education that undergraduate students receive in biology course labs. To this end, the group hosts an annual workshop/conference and produces the newsletter *Labstracts*. Members also receive copies of any previous proceedings.

Association for Women in Science (AWIS)

1200 New York Avenue, Suite 650
Washington, DC 20005
telephone: 202-326-8940
fax: 202-326-8960

e-mail: awis@awis.org
URL: http://www.awis.org/

The purpose of the Association for Women in Science is to help achieve "equity and full participation of women in all areas of science and technology." In existence for nearly three decades, the AWIS has 76 chapters in 42 states, and 5,000 national and international members. Membership is open to anyone who supports women in science.

The AWIS runs a comprehensive mentoring project and publishes information about such topics as grant availability, along with its quarterly *AWIS Magazine*. Each issue of the magazine carries news, job listings, and articles.

Committee for the National Institute for the Environment (CNIE)
1725 Kay Street, NW
Washington, DC 20006-1401
telephone: 202-530-5810
fax: 202-628-4311
e-mail: cnie@cnie.org
URL: http://www.cnie.org/

This committee is proposing the establishment of a National Institute for the Environment "to improve the scientific basis for making decisions on environmental issues." The CNIE began to form in 1990 when a group of scientists and business, political, and community leaders came together to discuss the decision-making process behind environmental issues. The CNIE eventually proposed the national institute, an idea that has garnered the support of more than 450 organizations.

Since then, a 1997 directive from the U.S. House of Representatives Appropriations Committee instructed the National Science Foundation to study the implementation and costs of a National Institute for the Environment. The National Science Foundation formed the National Science Board (NSB) Task Force on the Environment, and on July 28, 1999, the NSB backed recommendations implementing "nearly all of the activities proposed for a National Institute for the Environment," according to the CNIE.

Conservation International (CI)
2501 M Street, NW, Suite 200
Washington, DC 20037
telephone: 202-429-5660 or 800-429-5660
fax: 202-887-0193
e-mail: webmaster@conservation.org
URL: http://www.conservation.org/

As its name implies, this field-based organization furthers conservation of biodiversity in endangered ecosystems. With an objective of basing its actions on science in addition to policy, economics, and community involvement, CI, which was founded in 1987, works to protect global diversity with projects in 27 countries on four continents.

Ecological Society of America (ESA)
2010 Massachusetts Avenue, NW, Suite 400
Washington, DC 20036
telephone: 202-833-8773
fax: 202-833-8775
e-mail: esahq@esa.org
URL: http://www.sdsc.edu/~ESA/

This 7,600-member organization was founded in 1915 to "stimulate sound ecological research; clarify and communicate the science of ecology; and promote the responsible application of ecological knowledge to public issues."

The Ecological Society of America publishes *Ecology, Ecological Applications, Ecological Monographs, The Bulletin of the Ecological Society of America,* and other publications. It also offers a variety of educational resources, including a brochure detailing careers for high school and college students and a series of fact sheets and reports on topical issues, such as invasive species, biodiversity, and global climate change.

National Association of Biology Teachers (NABT)
12030 Sunrise Valley Drive, Suite 110
Reston, VA 20191-3409
telephone: 703-264-9696 or 800-406-0775
fax: 703-264-7778
e-mail: nabter@aol.com
URL: http://www.nabt.org/

More than 7,500 educators are members of the National Association of Biology Teachers, a group that "empowers educators to provide the best possible biology and life science education for all students." The association offers its members subscriptions to the journal *The American Biology Teacher* and to the newsletter *News & Views*. Published nine times a year, *The American Biology Teacher* tackles social and ethical issues, provides laboratory, classroom, and field teaching suggestions, plus reviews of books, software, and other educational media. The quarterly newsletter focuses on professional opportunities, teaching resources, NABT activities, and other news.

Each fall the association holds a national convention, which includes a variety of workshops, speakers, field trips, and presentations of research papers. Other NABT seminars and workshops run at various times and locations throughout the year.

National Audubon Society
700 Broadway
New York, NY 10003
telephone: 212-979-3000
fax: 212-979-3188
e-mail: webmaster@list.audubon.org
URL: http://www.audubon.org/

Founded in 1905, the National Audubon Society has more than half a million members, along with a full-time staff of scientists and educators, and managers for its 100 sanctuaries and nature centers. The society strives to "conserve and restore natural ecosystems, focusing on birds and other wildlife for the benefit of humanity and the Earth's biological diversity." Ornithologist and artist John James Audubon (1785–1851) is the society's namesake. The society publishes *Audubon*, a bimonthly magazine.

National Wildlife Federation (NWF)
8925 Leesburg Pike
Vienna, VA 22184
telephone: 703-790-4000
URL: http://www.nwf.org/nwf/

This large and well-known organization has roles as educator, activist, and advocate on issues ranging from endangered habitats and land stewardship to water quality and conservation. In education, the National Wildlife Foundation runs programs to encourage environment-friendly landscaping methods at home, work, and school; to make resources available for campus students, faculty, and administrators; to promote environmental knowledge and outdoor skills among children, and to foster family involvement in the out-of-doors. Other programs provide conservation materials for the classroom.

The National Wildlife Foundation also produces the bimonthly magazine *National Wildlife* and the children's magazines *Ranger Rick* and *Your Big Backyard*.

The Nature Conservancy (TNC)
International Headquarters
4245 North Fairfax Drive, Suite 100
Arlington, VA 22203-1606
telephone: 703-841-5300
URL: http://www.tnc.org/

The Nature Conservancy preserves land and waters around the world through arrangements with donors and sellers. As a measure of its success, the organization points to its U.S. nature sanctuaries, which number more than 1,500 and total some 10.5 million acres. Through donations, membership fees, and other support mechanisms, the conservancy continues to purchase land and operate its preserves. As of December 1998, the conservancy counted 900,000 members.

Sigma Xi, The Scientific Research Society
99 Alexander Drive
P.O. Box 13975
Research Triangle Park, NC 27709
telephone: 800-243-6534 or 919-549-4691
fax: 919-549-0090
URL: http://www.sigmaxi.org/

Since its beginnings in 1886 as an honor society for science and engineering, Sigma Xi has grown into an international research society with 80,000 members in more than 500 chapters. The society now also supports research, encourages public appreciation for research, and promotes international exchange between the public and the scientific/ technological sectors. To become an associate member, a person must show promise as a researcher, then receive an invitation to join. Noteworthy research warrants the status of full member.

Sigma Xi also publishes the bimonthly *American Scientist*, which covers biology as well as other disciplines. The society holds annual meetings and forums of current topics, runs grants-in-aid and distinguished lectureship programs, and offers a variety of prizes and awards.

Student Conservation Association (SCA)
689 River Road
P.O. Box 550
Charlestown, NH 03603-0550
telephone: 603-543-1700
fax: 603-543-1828
URL: http://www.sca-inc.org/

Every year, the Student Conservation Association gives 5,000 students opportunities to preserve and protect America's natural resources, while providing more than a million hours of service to national parks, wildlife refuges, and other natural areas. In essence, the group matches those seeking education, leadership, and personal development with various agencies and organizations that need help. The Student Conservation Association compiles information about seasonal and year-round volunteer, internship, and employment opportunities.

World Conservation Union (IUCN)
28 rue Mauverney
Ch-1196 Gland
Switzerland
telephone: 41 22 999-0001
fax: 41 22 999-0002
URL: http://www.iucn.org/

Begun in 1948 as the International Union for the Protection of Nature, the World Conservation Union is one of the oldest such organizations. The union is most commonly called the IUCN, which refers to a previous name of the organization. The union's membership comprises 74 governments and 105 governmental agencies, and more than 700 non-governmental organizations. Some 8,000 volunteer scientists and other experts help the IUCN develop and implement conservation strategies through six global commissions: The Commission on Ecosystem Management, The Commission on Education and Communication, The Commission on Environmental Economics and Social Policy, The Commission on Environmental Law, The Species Survival Commission, and The World Commission on Protected Areas.

The IUCN generates the Red List of Threatened Species, and makes available to its members and the general public a wide range of information on conservation practices and legislation.

World Wide Fund for Nature (WWF)
1250 24th Street, NW
Washington, DC 20037
telephone: 800-225-5993
URL: http://www.worldwildlife.org/

Founded in 1961, the World Wide Fund for Nature has more than 5 million supporters and more than two dozen national and associate organizations around the world. The organization's goals are to protect biological diversity at the genus, species, and ecosystem levels—with a particular emphasis on wetlands and coasts, along with forests and woodlands; to control pollution; and to advance the sustainable use of natural resources. Besides extensive fieldwork, the WWF also has a strong education component and works to increase public awareness of various topics of interest.

SPECIALIZED GROUPS

American Association of Anatomists (AAA)
9650 Rockville Pike
Bethesda, MD 20814-3998
telephone: 301-571-8314
fax: 301-571-0619
e-mail: exec@anatomy.org
URL: http://www.anatomy.org/anatomy

The American Association of Anatomists dates back more than a century, to 1888. It has become a highly respected organization with a membership that has included more than 100 members of the National Academy of Sciences. Current members of this multidisciplinary group represent a broad spectrum of fields, but all share an interest in morphology on its many levels from microscopic to macroscopic. AAA members and those of other organizations attend the annual Experimental Biology meeting, which draws speakers from various fields.

The association also has two scientific journals, *Developmental Dynamics* and *The Anatomical Record,* and publishes a quarterly newsletter. Through these publications and other efforts, the group takes an active role in educational and legislative issues.

American Fisheries Society (AFS)
5410 Grosvenor Lane, Suite 110
Bethesda, MD 20814-2199
telephone: 301-897-8616
fax: 301-897-8096
e-mail: main@fisheries.org
URL: http://www.fisheries.org/

This large, professional organization serves fisheries scientists and "promotes scientific research and enlightened management of resources for optimum use and enjoyment by the public." Founded in 1870, the American Fisheries Society publishes several of the premier publications in the field, including *Transactions of the American Fisheries Society, North American Journal of Fisheries Management, North American Journal of Aquaculture; The Journal of Aquatic Animal Health,* and *Fisheries.*

In addition to the organization of a variety of meetings, the society is involved with professional certification, international affairs, and other areas.

Members of the AFS can participate in one of the society's 54 chapters and student subsections, its four regional divisions, and dozens of committees. Special-interest sections on such topics as early life history and estuaries are also available to members.

American Ornithologists' Union (AOU)
c/o Division of Birds
MRC 116
National Museum of Natural History
Washington, DC 20560
URL: http://pica.wru.umt.edu/AOU/AOU.html

The American Ornithologists' Union has grown from its origination in 1836 into a major organization for the scientific study of birds. While the union is primarily directed at professionals, its membership of 4,000 contains many amateurs.

The union publishes the quarterly journal *The Auk*, which presents original research and book reviews, in addition to other items of interest. It also publishes a series called *Ornithological Monographs* for longer research papers, a bimonthly newsletter, and the *Check-List of North American Birds*.

American Society for Biochemistry and Molecular Biology (ASBMB)
9650 Rockville Pike
Bethesda, MD 20814-3996
telephone: 301-530-7145
fax: 301-571-1824
e-mail: asbmb@asbmb.faseb.org
URL: http://www.faseb.org/asbmb/

A scientific and educational organization, the American Society for Biochemistry and Molecular Biology dates back to 1906. It has more than 10,000 members, most of whom are researchers at institutions of higher learning or in government, industry, or nonprofit laboratories, or are teachers at colleges and universities. The society produces *The Journal of Biological Chemistry* and other publications.

American Society for Horticultural Science (ASHS)
600 Cameron Street
Alexandria, VA 22314-2562
telephone: 703-836-4606
fax: 703-836-2024

e-mail: webmaster@ashs.org
URL: http://www.ashs.org/

The American Society for Horticultural Science has been advancing research and education within the field since its inception in 1903. Its 4,500 members represent a variety of specialties and include plant science researchers, government-run experimental stations, extension agencies, commercial growers, and others. The society holds an annual international conference and other meetings, and runs various educational programs. In addition to ASHS scientific journals and a monthly newsletter, the society's press publishes books for horticulturists.

American Society for Microbiology (ASM)
1752 N Street, NW
Washington, DC 20036-2804
telephone: 202-737-3600
e-mail: membership@asmusa.org
URL: http://www.asmusa.org

The American Society for Microbiology, which is open to interested individuals with a bachelor's degree or equivalent in the field, has more than 43,000 members from around the world. The ASM caters to two dozen specialty disciplines and has a separate division for educators in the field.

The society produces 10 journals, including the *Journal of Bacteriology, Molecular and Cellular Biology*, the *Journal of Clinical Microbiology*, and the *Journal of Virology*. The society also runs the ASM Press and organizes a number of meetings and workshops each year. For students, it offers information about microbiological career opportunities and other educational resources, travel grants, a limited number of predoctoral fellowships, and other awards.

American Society of Ichthyologists and Herpetologists (ASIH)
Robert Karl Johnson, Secretary
Grice Marine Laboratory
University of Charleston
205 Fort Johnson Road
Charleston, SC 29412
telephone: 843-406-4017
fax: 843-406-4001
e-mail: johnsonr@cofc.edu
URL: http://www.utexas.edu/depts/asih/

The American Society of Ichthyologists and Herpetologists gives its attention to the scientific study of—and dissemination of information about—fish, amphibians, and reptiles. Through its programs and its 2,400 members, the society also contributes to the global understanding of biodiversity and the need to conserve it. In addition, a volunteer committee program highlights topics of interest, including animal welfare, biodiversity and conservation, environmental quality, nomenclature, scientific and public education, and scientific research collections, resources, and practices.

This 85-year-old organization produces *Copeia*, a quarterly journal named for 19th-century American scientist Edward Drinker Cope, and special publications based on symposia held during its annual meetings. These meetings draw scientists and students from around the world.

American Society of Limnology and Oceanography (ASLO)
Asit Mazumder, ASLO Secretary
Department of Biological Science
University of Montreal
C.P. 6128, Succ. A Quebec
Montreal, Quebec H3C 3J7
Canada
telephone: 514-343-2286
fax: 514-343-2293
e-mail: secretary@aslo.org
URL: http://www.aslo.org/

Formed by the 1948 merger of two groups—the Limnological Society of America and the Oceanographic Society of the Pacific—the American Society of Limnology and Oceanography promotes research and discussion of aquatic-science topics and has become involved with education, policy-making issues, and scientific ethics. With some 4,000 members from around the world, the society holds annual, interdisciplinary meetings and special symposia. It also produces the widely cited journal *Limnology and Oceanography* and the more informal, member-oriented *ASLO Bulletin*.

American Society of Naturalists
c/o The University of Chicago Press, Journals Division
P.O. Box 37005
Chicago, IL 60637
telephone: 773-753-3347
fax: 773-753-0811
e-mail: subscriptions@journals.uchicago.edu
URL: http://www.amnat.org/

Established in 1883, the American Society of Naturalists took over the publication of *American Naturalist*, which dates back to 1867. Both

American Naturalist, a monthly journal, and the society focus on evolution, adaptation, community and ecosystem dynamics, ecology, and other broad biological concepts. The society also holds annual meetings and offers several awards, including the E. O. Wilson Naturalist Award. Membership is open to those who have continued to publish scholarly articles in the field.

American Society of Plant Physiologists (ASPP)
15501 Monona Drive
Rockville, MD 20855-2768
telephone: 301-251-0560
fax: 301-279-2996
e-mail: knoone@aspp.org
URL: http://www.aspp.org/

Since the American Society of Plant Physiologists began in 1924, it has expanded its role from promoting plant physiology and research and now also encompasses related molecular and cellular biology interests. Despite the society's name, its members come from around the world. The society offers two monthly journals, *Plant Physiology* and *The Plant Cell*, and the bimonthly *ASPP Newsletter*. Other society services include an annual meeting, sectional societies, a public affairs program, a job bank, and the ASPP Education Foundation.

American Society of Plant Taxonomists (ASPT)
ASPT Business Office
Department of Botany
University of Wyoming
Laramie, WY 82071-3165
telephone: 307-766-2556
fax: 307-766-2851
e-mail: aspt@uwyo.edu or ljbrown@uwyo.edu
URL: http://www.sysbot.org

Research and teaching of the taxonomy, systematics, and phylogeny of vascular plants is the focus of the 1,400-member American Society of Plant Taxonomists. The ASPT, which was organized in 1935, publishes *Systematic Botany*, a quarterly journal, along with an electronic newsletter and *Systematic Botany Monographs*, and also holds annual meetings.

Animal Behavior Society (ABS)
American Editorial Office for Animal Behaviour
2611 East 10th Street, #170
Indiana University
Bloomington, IN 47408-2603

telephone: 812-856-5541
fax: 812-856-5542
e-mail: aboffice@indiana.edu
URL: http://www.animalbehavior.org/ABS/

Dating back to 1964, the Animal Behavior Society fosters the biological study of animal behavior, including everything from hibernation and foraging to courtship and mating. The society holds annual meetings that bring together scientists from the United States and beyond. It also publishes a journal, *Animal Behaviour*.

Aquatic Ecosystem Health and Management Society

P.O. Box 85388
Brant Plaza Postal Outlet
Burlington, Ontario L7R 4K5
Canada
fax: 905-634–3516
URL: http://www.aehms.org/

This society, which began in 1989, has as its main mission the promotion and development of "the concept of ecosystem health for the protection, conservation, and sustainable management of global aquatic resources." It is also involved in remediation and restoration of aquatic ecosystems. The society holds biennial conferences and organizes various topic-driven symposia, and has a series of specialty working groups. Its publications include the journal *Aquatic Ecosystem Health and Management*.

Association for the Study of Animal Behaviour (ASAB)

The Membership Secretary
82A High Street
Sawston, Cambridge, CB2 4HJ
United Kingdom
e-mail: asab@grantais.demon.co.uk
URL: http://www.hbuk.co.uk/ap/asab/

Like the Animal Behavior Society, the Association for the Study of Animal Behaviour advances the study of animal behavior. Its 1,800 members are primarily from Britain and other areas of Europe. While most are professional biologists or educators, the association's membership is open to anyone with an interest in animal behavior.

The group also holds several meetings and workshops each year and publishes (with the Animal Behavior Society) the monthly international publication *Animal Behaviour*, which includes reviews, papers, research articles, and book reviews. Research areas covered in the publication include: behavioral ecology, the evolution of behavior, ethology, population biology, and navigation and migration. In addition, the society publishes the *ASAB Newsletter* three times each year to inform its members of job openings, upcoming conferences, and other news.

Association of Field Ornithologists (AFO)
c/o Allen Press
P.O. Box 1897
Lawrence, KS 66044-8897
URL: http://www.afonet.org/index.html

Following its historical origins as the New England Bird Banding Association and the Northeastern Bird-Banding Association, the Association of Field Ornithologists continues to be active in bird-banding and field techniques but has expanded to include other types of scientific study. The association promotes research activities among both its professional and amateur members and encourages conservation biology.

The association holds annual meetings and publishes the quarterly *Journal of Field Ornithology*, which includes articles on field techniques, life history, and bird distribution, ecology, and behavior. It has recently emphasized the birds of the neotropics. The AFO also publishes the annual *Bird Count Supplement*, the bimonthly *Ornithological Newsletter*, and the periodic newsletter *AFO Afield*.

Botanical Society of America (BSA)
1735 Neil Avenue
Columbus, OH 43210-1293
telephone/fax: 614-292-3519
e-mail: srussell@ou.edu
URL: http://www.botany.org/

The Botanical Society of America advances teaching and research in all areas of plant biology from reproductive biology to evolution, and from systematics to ecology. Its 15 special-interest sections provide additional resources for members. It publishes the monthly *American Journal of Botany* along with the informal quarterly *Plant Science Bulletin*.

The society organizes annual and regional meetings and offers travel grants to the International Botanical Congress, which is held once every six years. The 1999 congress was held in the United States. Students can also find career information through the Botanical Society of America.

The Crustacean Society
Business Office
P.O. Box 1897

Lawrence, KS 66044-8897
telephone: 785-843-1221
fax: 785-843-1274
e-mail: jeff@vims.edu
URL: http://www.lam.mus.ca.us/~tcs/

This fairly new society is an outgrowth of the Crustacean Club, a less
formal group of biologists who discussed the idea of forming a society
during the annual meeting of the American Society of Zoologists in 1977.
The club voted in 1979 to form The Crustacean Society, which sought its
first members the following year.

The society encourages research and information exchange on the
topics of fossil and recent crustaceans through its annual meeting, various
symposia and workshops, awards for students and researchers, and its
Journal of Crustacean Biology.

Declining Amphibian Populations Task Force (DAPTF)
Office of Biodiversity Programs
NHB Mail Stop 180
Smithsonian Institution
Washington, DC 20560
telephone: 202-357-2620
fax: 202-786-2934
e-mail: Heyer.Ron@nmnh.si.edu
URL: http://www2.open.ac.uk/Ecology/J_Baker/JBtxt.htm

The Declining Amphibian Populations Task Force got its start in 1991
following reports of worldwide losses in amphibian populations, and in
some cases, species extinctions. Based at the Open University Ecology and
Conservation Research Group and operating in association with the
IUCN (World Conservation Union) Species Survival Commission, the
DAPTF hopes "to determine the nature, extent and causes of declines of
amphibians throughout the world, and to promote means by which
declines can be halted or reversed." The organization has more than
3,000 scientists and conservationists associated with it through its working groups.

The task force produces the periodic newsletter *Froglog*, which provides updates on research along with information about task force research funding. The newsletter can be viewed on the DAPTF Web site.

The Endocrine Society
4350 East West Highway, Suite 500
Bethesda, MD 20814-4410
telephone: 301-941-0200

fax: 301-941-0259
e-mail: endostaff@endo-society.org
URL: http://www.endo-society.org/

The Endocrine Society is an organization with an international member-ship of more than 9,000 researchers, educators, medical professionals, and students. Society members are interested in basic and applied endo-crinology research, and clinical applications that relate to the body's hormone-producing glands and tissues. In addition to advancing the work of its members, the society is also involved with enhancing the public awareness and appreciation for the endocrine system and for endocrine research.

The Endocrine Society holds an annual meeting and runs various courses and conferences during the year. It also publishes the journals *Endocrinology*, *The Journal of Clinical Endocrinology and Metabolism*, *Endocrine Reviews*, and *Molecular Endocrinology*.

Entomological Society of America (ESA)

9301 Annapolis Road
Lanham, MD 20706-3115
telephone: 301-731-4535
fax: 301-731-4538
e-mail: esa@entsoc.org
URL: http://www.entsoc.org/

This society's 7,400 members include scientists, researchers, teachers, and consultants with a wide variety of affiliations. The Entomological Society, which is actually the result of a 1953 merger of the American Association of Economic Entomologists and the Entomological Society of America, promotes the science and profession of entomology. Its publica-tions include *American Entomologist*, *Annals of the ESA*, *Environmental Entomology*, the *Journal of Economic Entomology*, and the *Journal of Medical Entomology*, among others.

Microscopy Society of America (MSA)

MSA Business Office
435 North Michigan Avenue, Suite 1717
Chicago, IL 60611-4067
telephone: 800-538-3672 or 312-644-1527
fax: 312-644-8557
e-mail: BusinessOffice@MSA.Microscopy.com
URL: http://www.msa.microscopy.org/

Founded in 1942 as The Electron Microscope Society of America, the since-renamed Microscopy Society of America has expanded its scope to encompass all areas of microscopy. The society focuses not only on instrumentation and techniques, but on the applications, development, and use of microscopes, and related images and analysis. With some 4,500 members, including about 500 students, the society is affiliated with 30 local and regional groups. The MSA also holds annual meetings, produces the *MSA Journal,* and offers a certification program for electron microscope technologists in biological sciences.

Organization of Biological Field Stations (OBFS)
P.O. Box 247
Bodega Marine Laboratory
Bodega Bay, CA 94923
e-mail: stromber@socrates.berkeley.edu or obfs@ucdavis.edu
URL: http://www.obfs.org/

Professionals from biological field stations in North and Central America make up the majority of the 200-strong membership of the Organization of Biological Field Stations. The organization produces a newsletter and other reports, and holds an annual meeting with the primary purpose of encouraging discussion and information exchange among field station directors.

In addition, the organization helps disseminate information about research opportunities for undergraduate students along with employment opportunities at various stations. The OBFS Web site also contains links to many individual field stations. These links provide course listings and other details about the programs and opportunities available at the stations.

Orthopterists' Society
J. A. Lockwood
Entomology Section
Department of Renewable Resources
University of Wyoming
Laramie, Wyoming 82071-3354
telephone: 307-766-4260
fax: 307-766-5025
e-mail: lockwood@uwyo.edu
URL: http://viceroy.eeb.uconn.edu/OS_Homepage

The Orthopterists' Society began in 1978 as the result of discussions between 50 orthopterists who were meeting in Argentina. It now has an international membership of 330 students and scientists who are interested in grasshoppers, crickets, and related insects. Society members' research projects encompass taxonomy, ecology, control measures, physiology, and other topics.

The society holds periodic international meetings, along with a symposium at each annual meeting of the Entomological Society of America. In addition to its semiannual newsletter *Metaleptea*, the "Proceedings of the Orthopterists' Society" appear as a special issue of the *Journal of Orthoptera Research*.

Raptor Research Foundation (RRF)
c/o OSNA
P.O. Box 1897
810 E. 10th Street
Lawrence, KS 66044-8897
URL: http://biology.boisestate.edu/raptor

This nonprofit scientific organization is devoted to the study of raptors, which include hawks, eagles, and other birds of prey. Its activities are both educational and conservation-oriented. More than 1,200 researchers, students, lay people, and others form the membership of the Raptor Research Foundation, which originated in 1966. The foundation also publishes a newsletter and a quarterly, *The Journal of Raptor Research*. In addition to occasional workshops and seminars, the Raptor Research Foundation holds an annual meeting to present some of the year's research findings.

Society for Conservation Biology (SCB)
Alice Blandin, Executive Coordinator
Society for Conservation Biology
University of Washington
Box 351800
Seattle, WA 98195-1800
telephone: 206-616-4054
fax: 206-543-3041
e-mail: conbio@u.washington.edu
URL: http://conbio.rice.edu/scb/

The Society for Conservation Biology is an international professional organization with a membership that includes students, educators, various conservationists, and others. Its particular emphasis is on scientific and technical studies concerned with the conservation and study of biological diversity. The society holds annual meetings, has a number of local chapters, and publishes the *Journal of Conservation Biology*.

Society for Developmental Biology (SDB)
9650 Rockville Pike
Bethesda, MD 20814-3998
telephone: 301-571-0647
fax: 301-571-5704
e-mail: sdb@faseb.org
URL: http://sdb.bio.purdue.edu/

Open to anyone possessing a doctoral degree and an interest in the field, and to students enrolled in a related graduate program, the Society for Developmental Biology is "dedicated to the advancement of scientific and educational excellence in developmental biology, and to the public's understanding of developmental biology." The society dates back more than six decades.

The Academic Press publishes *Developmental Biology* under the auspices of the society. The Society for Developmental Biology also holds a major annual meeting, along with several regional meetings. Through its Web site, SDB offers information about, and links to, developmental-biology resources.

Society for Integrative and Comparative Biology (SICB)
401 N. Michigan Avenue
Chicago, IL 60611-4267
telephone: 312-527-6697 or 800-955-1236
fax: 312-527-6705
e-mail: scib@sba.com
URL: http://www.sicb.org/

The Society for Integrative and Comparative Biology formed in 1902 when the Central Naturalists and the American Morphological Society merged. The society strives to help advance and develop the field of biology, in part by integrating the many areas of specialization within the field and by disseminating important biological information to the public. Through the organization, biologists share their research in the society's journal, *American Zoologist*, and by presentations at the society's many symposia, including its annual meetings.

The SICB has 11 divisions. They are: animal behavior, comparative endocrinology, comparative physiology and biochemistry, developmental and cell biology, evolutionary developmental biology, ecology and evolution, integrative and comparative issues, invertebrate zoology, neurobiology, systematic and evolutionary biology, and vertebrate morphology.

In all, the Society for Integrative and Comparative Biology has 2,100 members.

Society for Neuroscience
11 Dupont Circle, NW, Suite 500
Washington, DC 20036
telephone: 202-462-6688
e-mail: info@sfn.org
URL: http://www.sfn.org/

The brain, spinal cord, and peripheral nervous system are the topics of interest to the 25,000 members and 100 chapters of the Society for Neuroscience. The society strives to bring together researchers, promote education and research in the field, and disseminate information to the public about recent findings and their implications. The fall annual meeting and the society's periodical, *The Journal of Neuroscience*, showcase new research in neuroscience. The society provides publications on the brain and neuroscience topics for use by the general public, and a newsletter for Congressional staff members.

Society for the Study of Amphibians and Reptiles (SSAR)
Robert D. Aldridge, Treasurer/Publications Secretary
Department of Biology
St. Louis University
3507 Laclede Avenue
St. Louis, MO 63103-2010
telephone: 314-977-3916 or 314-977-1710
fax: 314-977-3658
e-mail: SSAR@slu.edu
URL: http://www.ukans.edu/~ssar/

Founded in 1958, the Society for the Study of Amphibians and Reptiles is one of the best-known societies within the field. It produces the often-cited, quarterly *Journal of Herpetology* and the "news-journal" *Herpetological Review*, and it also publishes a variety of books and other publications. The society offers grants and provides information about grant opportunities the world over.

The SSAR encourages student and international participation in its an annual, weeklong meeting, which is held each summer on a university campus. The meeting includes various symposia, workshops, and field trips.

Society for the Study of Evolution
Business Office
P.O. Box 1897
Lawrence, KS 66044-8897
telephone: 800-627-0932
URL: http://lsvl.la.asu.edu/evolution/

Since 1946, the Society for the Study of Evolution has been advancing the interdisciplinary study of "organic evolution." It holds meetings and publishes the bimonthly *Evolution*. Members of the society include students, researchers, and anyone with an interest in evolution.

Society of Wetland Scientists (SWS)
Business Office
P.O. Box 1897
Lawrence, KS 66044-8897
telephone: 785-843-1221 or 800-627-0629
fax: 785-843-1274
e-mail: sws@allenpress.com
URL: http://www.sws.org/

Established in 1980, the Society of Wetland Scientists fosters the understanding, conservation, and knowledgeable management of wetlands. Besides providing a forum for information exchange about wetlands among scientists, the society takes an active role in educating the public about the importance of these ecosystems.

The society has more than 4,000 members from North American and other continents. An annual meeting gives scientists a chance to learn about wetland information and research, and other society programs provide support for wetland science and student research. Members of the Society of Wetland Scientists receive the quarterly *Wetlands* journal, the *SWS Bulletin*, and chapter newsletters. *Wetlands* is an international journal "with the goal of centralizing the publication of pioneering wetlands work that is otherwise spread among a myriad of journals."

Wilson Ornithological Society or The Wilson Society
Museum of Zoology
University of Michigan
1109 Geddes Avenue
Ann Arbor, MI 48109-1079
telephone: 734-764-0457
e-mail: jhinshaw@umich.edu
URL: http://www.ummz.lsa.umich.edu/birds/wos.html

The worldwide Wilson Society is an ornithological group of nearly 2,500 professional and serious amateur birders. Headquartered in the Bird Division of the University of Michigan Museum of Zoology, the society holds annual meetings and publishes the quarterly journal *The Wilson Bulletin* and a bimonthly *Ornithological Newsletter*. Professionals and amateurs alike can take advantage of some of the many paid and volunteer position openings listed in the newsletter. Students may find excellent field opportunities among the postings.

CHAPTER THIRTEEN
Glossary

acid rain   Precipitation with a pH of less than 5.6. Acid rain is often the result of a reaction between water and atmospheric sulfur dioxide or various nitrogen oxide gases. These gases can be released from factories and coal-fired power plants, car emissions, lightning, forest fires, and volcanoes.

alleles   Different forms of a single gene.

amino acid   A building block in a *protein*, an amino acid consists of an *amino group* (NH_2) and a carboxyl group (COOH) linked to one carbon atom, along with a side chain, or R group. The R group distinguishes one amino acid from another. Only 20 of the 80-plus naturally occurring amino acids are commonly found in proteins.

amino group   The NH_2 component of an *amino acid*.

anaerobic   Anaerobic organisms do not need oxygen to survive. Obligate anaerobes cannot survive in the presence of oxygen; facultative anaerobes can survive whether oxygen is present or not.

anapsid   Reptiles classified as anapsids, such as turtles, lack temporal fossa (openings in the temple region of the skull, behind the eyes). (See also *diapsid*.)

angiosperm   A plant that has its seeds in an ovary. Angiosperms are also known as flowering plants.

annelid   A segmented worm, such as an earthworm or a leech.

anthropological genetics   A new field, practitioners of which study the geographical origin of humans.

antibiotic   A substance produced by a microorganism—or synthesized in a lab—that adversely affects other microorganisms.

archaea   Microbes that live in extreme environments, such as deep-sea hot springs or areas of high salt or sulfur concentrations. Although archaea look like *bacteria*, genetic analyses have placed them nearer the *eukaryotes*. (See *prokaryote*.)

arthropod   An organism with jointed limbs, such as an insect, spider, or crustacean.

artiodactyl   An ungulate (herbivorous, hoofed animal) that has an even number of toes. Artiodactyls include cows, pigs, and deer.

asteroid   An astronomical term for the small, planet-like celestial bodies orbiting the Sun. Most are located between Mars and Jupiter.

astrobiology   The scientific study of extraterrestrial life.

bacteria   *Prokaryotic* organisms that are usually unicellular. Bacteria can be spherical (spirilli), helical (cocci), or rod-shaped (bacilli). Some are beneficial to humans, while others can cause disease.

base   A building block of DNA and RNA. A base can be either a single-ringed pyrimidine, like cytosine, thymine, and uracil, or a double-ringed purine, like adenine or guanine. When combined with a sugar and a phosphate group, they form nucleotide bases that link together to form DNA.

base-pair   A complementary pairing of nucleotide *bases*, seen in DNA. Pairing also occurs during *protein* synthesis between DNA and the form of RNA called messenger RNA. DNA's four bases—adenine, thymine, guanine, and cytosine—pair up as adenine-thymine and guanine-cytosine. During protein synthesis, the same pairing occurs, except that RNA contributes uracil as a base instead of thymine.

bilateral symmetry   Animals with bilateral symmetry have near-mirror-image right and left sides, plus a head and rear. (See *radial symmetry*.)

biodiversity   The variety of living things—and associated ecological processes—in a particular area or region. It is also used as a general term referring to the diversity of all life on Earth.

bio-indicator   An organism that in some way provides information about the health of an *ecosystem*. The presence of certain frogs, for instance, may indicate that a wetland is relatively free of specific pollutants.

biological clock   An internal time-keeping mechanism that helps direct the daily, seasonal, and/or annual activity of animals and plants. Many organisms use a biological clock in conjunction with environmental cues to regulate their activities.

biology   The scientific study of living things.

bioluminescence   The production of light (without heat) by various organisms, such as fireflies and many deep-sea animals.

biomass   The mass of all organisms—or of all those under investigation within a certain study area or at a particular *trophic level*. For example, the biomass of amphibians in a wildlife refuge is the total mass of all frogs, salamanders, and caecilians within that area.

biovolume The volume of an organism. For example, a newly discovered giant sulfur bacterium has up to 100 times the biovolume of other *prokaryotes*.

biphasic A characteristic of organisms that live one part of their lives very differently from another. Many amphibians, for example, have an aquatic larval stage and a terrestrial adult stage.

botany The scientific study of plants.

branch (re: tree of life) A grouping of organisms with similar characteristics. *Evolution*ary biologists often group organisms into branches on a *tree of life* based on such factors as *morphological* similarities and fossil evidence. Branches can be broadly defined, such as a branch that includes all *prokaryotes*, or more narrowly defined, such as separate branches for apes and chimpanzees.

Cambrian explosion The Cambrian geological period, which began about 590 to 570 million years ago and lasted until 500 to 505 million years ago, was one of the most productive in Earth's history in terms of the diversification of animals.

carrying capacity The maximum number of individuals in a selected population that a particular environment can sustainably support.

centromere The connection point for the two halves of a *chromosome*. In a *eukaryote*, the two halves, called chromatids, separate as a step in cell division.

character (re: cladistics) A trait exhibited by a *species* or group of species. Biologists who use *cladistics* in their studies of systematics group organisms according to the presence or absence of specific characters.

chemosynthesis The use by living organisms of chemicals for their energy (as opposed to *photosynthesis*, in which living things rely on sunlight for energy).

chordate A member of the phylum Chordata. The phylum encompasses a large group of animals that includes the vertebrates and the protochordates (such as tunicates). Chordates all have a notocord, a flexible support structure that runs the length of the body (in many vertebrates, the vertebral column has essentially replaced the notocord); a dorsal hollow nerve cord; and gill slits at some point in development.

chromatin Within the cell *nucleus*, the *chromatin* is a complex of *DNA* and DNA-binding *protein*, including the histones.

chromosome Located within the cell *nucleus*, chromosomes contain the *genes*. Human *somatic cells*, which are all but the sperm and (unfertilized) egg cells, have 46 chromosomes. Fruitflies have eight.

CITES Convention on International Trade in Endangered Species of Wild Fauna and Flora. CITES lists various *species* that are threatened with *population* declines and are regulated for international trade.

cladistics A method of classifying living things that relies upon differences and similarities between the various *characters* of organisms, and ancestral relationships. A clade is a hypothetical grouping that indicates similar (and possibly unique) characters of the organisms therein.

clone A genetically identical duplicate.

contemporaneous When two *species* live during the same time period, the two are said to be contemporaneous.

coprolite Specimen of fossilized dung.

creationism Belief that the *species* on Earth were created at one time and in their fully formed states by a divine being or force. Creationism rejects the scientifically based theory of *evolution*.

Darwin, Charles R. (1809–82) Naturalist who collaborated with Alfred R. Wallace on the 1858 paper that explained their ideas about *evolution*. A year later, Darwin published a book, *On the Origin of Species by Means of Natural Selection, or the Preservation of Favoured Races in the Struggle for Life*. The paper and book initiated discussion and scientific study of the theory of evolution, which has since become the foundation of the biological sciences. (See *natural selection*.)

DDT Dichlorodiphenyl-tri-chloroethane. Banned in 1972, this insecticide was widely used in the United States following World War II. It became particularly concentrated in birds of prey, which experienced precipitous population declines due to the thinning of their eggshells. The shells became so thin that they cracked under the weight of the brooding parent.

deuterostome A division in animal classification that refers to a set of embryonic *characters*, especially how the mouth develops. Deuterostomes include the *chordates* (higher animals).

diapsid Reptiles classified as diapsids have openings (fossa) in the temporal area on each side of the skull. Diapsids include such reptiles as crocodiles, lizards, and snakes. (See *anapsid*.)

dicot A flowering plant that has two cotyledons (initial leaves). (See *monocot*.)

DNA Deoxyribonucleic acid (DNA) is the genetic material of nearly all organisms. (See *RNA*.) DNA contains sugar (deoxyribose), along with purine and pyrimidine *bases*. DNA has a double-helical shape, which appears somewhat like a twisted ladder that has stacked sugar molecules forming the main supports and bases forming the rungs.

DNA chip A DNA chip, or array, looks like a computer chip but has strips of *DNA* instead of transistors. DNA chips are designed to detect whether a cell or a *virus* exactly matches one of its tens of thousands of different DNA strips.

ecology The scientific study of living things and their relationships to the living and nonliving environment.

ecosystem The combination of all of the living and nonliving parts of a particular environment.

embryonic cells Cells of the embryo, representing the stage of life between conception and birth.

endangered A term applied to living things that are threatened with *extinction*.

endemic species Animals or plants that are native to only one particular area, which may be small or large.

environment An organism's external surroundings, which include living and nonliving components. Nonliving components can include such things as climate or soil and water conditions.

enzyme A *protein* that acts as a biological catalyst. Enzymes are usually very specific in their targets and functions.

eukaryote An organism with a membrane-clad *nucleus* and cellular organelles. (See *prokaryote*.)

Europa A moon of Jupiter.

evolution Change in living *populations* over time, with current species evolving from more ancient forms via the process of *natural selection*. The theory of evolution is widely accepted in the scientific community. (See *Darwin, Charles.*)

extant Term used for *taxa* that are currently living. For example, bald eagles are extant, but at one time they were on the brink of *extinction*.

extinction The loss of an entire *taxon* from Earth. For example, the passenger pigeon (*Ectopistes migratorius*), which once was so numerous that flocks could darken the sky, was hunted to extinction in the twentieth century.

extirpation The loss of a *population* of a *species* from a certain area on Earth. For example, a population of frogs might become extirpated from a specific area if the area's only pond were drained.

food chain The food chain is a depiction of the dietary connections among living things. As the food of most other organisms, plants are called primary producers and usually are placed at the bottom of the food chain. Herbivores, the organisms that feed on plants, usually make up the next step, also called a *trophic level*. Carnivores that feed on the herbivores, then, would represent a trophic level above herbivores. An example of a typical food chain would follow the order: lettuce (primary producer)—rabbit (primary consumer)—wolf (secondary consumer).

forb Broad-leaved herbaceous plant.

fungus A plant-like organism that lacks chlorophyll, which permits photosynthesis. Fungi obtain their energy instead from dead or decaying organic materials, or via a parasitic relationship with a host organism. (See *parasite*.)

gastric system The anatomy and biological processes pertaining to the stomach.

gene The basic unit of heredity. Each gene contains a sequence of *DNA* (or in a *virus*, *RNA*) that contains the code for one or more specific functions. Different forms of a single gene are called *alleles*.

genetic code A collection of *DNA bases*. The genetic code is read three bases at a time, and each of these triplets, or codons, relates to one *amino acid*. *Protein*s form when amino acids link together. With four DNA bases, the number of possible triplets is 64. Because several triplets can relate to the same amino acid, only 20 amino acids are commonly found in proteins.

genetic engineering The exchange of *DNA* to form a new, so-called *recombinant organism*. In its broadest definition, genetic engineering can refer to *selective breeding*, but the term is primarily reserved to describe the introduction and insertion of genetic material from one organism into another different organism.

genetic enhancement The improvement of an organism by means of *genetic engineering*. The term is usually applied to the potential future manipulations of a human's *genome* to enhance such traits as intelligence or athletic prowess.

genetic letter Colloquial term for one of the four *bases* in *DNA* or in *RNA*.

genetics The science surrounding the study of *genes*, the "tools" of heredity and variation.

genome The term is applied to the complement of *genes* in a *species* or the specific genome of an individual within a species.

genome patent Genome patents do not actually patent the *genome*. Instead, they patent the process used to generate the *genetic code* of an organism, along with the resulting database. Others can then access the patented database (such access would likely carry a fee) or generate the database themselves using an unpatented process.

genome sequencing The reading of an organism's full *genetic code*.

genus A grouping of *species* believed to be closely related. In biology's standard binomial nomenclature, the name of the genus is capitalized and precedes the lowercased species name, e.g., *Rana pipiens.* Both names are italicized.

Geographical Information System (GIS) Technology that links a vast, combined database geographically, or spatially, to Earth. GIS brings together already-collected, digital information about an area, allowing a biologist, for example, to select certain data, overlay it on a map, and then conduct analyses that are pertinent to his or her specific research project.

global warming A planet-wide average increase in temperature. Biologists and other scientists are currently debating the extent of the increase and its long-term effect(s) on life.

greenhouse gas Greenhouse gases include carbon dioxide, nitrous oxide, methane, ozone, and several classes of fluorine-, chlorine- and bromine-containing halocarbons. These gases trap solar radiation, which heats the Earth's atmosphere. Many scientists believe the gases are a cause of *global warming*.

gymnosperm Gymnosperms are the conifers and related species. They include a variety of mainly evergreen, cone-bearing trees and shrubs. Examples are the pines, spruces, and cedars. Larches, which are deciduous, are also gymnosperms.

habitat An animal's or plant's home environment. For example, a frog's habitat may be a wetland, whereas a corn-boring beetle may call a farm field its home.

haplotypes Different versions of a gene.

hectare 10,000 square meters, or the equivalent of 2.471 acres.

herpetology The scientific study of amphibians and reptiles.

higher plants Flowering plants and grasses.

hormone Chemical substances produced by plants and animals that, sometimes even in small quantities, act as chemical messengers and control responses within the organism.

host eukaryotic cell   This term refers to a *eukaryotic* cell that serves as host (temporary home) for viruses, which can then reproduce. Researchers also use host cells in cloning by inserting foreign *DNA*, which can then replicate.

human artificial chromosome (HAC)   A laboratory-generated human *chromosome*. Human artificial chromosomes hold promise as carriers of selected *genes* into human cells.

hydrothermal vents   Underwater springs that spew heated water from the sea floor.

ichthyology   The scientific study of fish.

intermediate disturbance (re: diversity)   A hypothesis about the cause of diversity in tropical rainforests: When a tree in the forest falls, the hypothesis suggests, the light gap created in the forest canopy creates conditions favorable for *pioneer species* to take hold.

introduced species   *Species* that are not native to the area in which they now live. Humans have both purposely and inadvertently transported and transplanted many non-native plants and animals to new *habitats* and even new continents. Introduced species often have unexpected and sometimes detrimental effects on native species.

keystone species   A *species* that is key to balancing an entire *ecosystem*. Sea otters along the coast of Alaska, for example, eat sea urchins, which eat kelp. When sea otter populations decline, sea urchins overpopulate and decimate kelp forests. In this scenario, the sea otter is a keystone species.

lithoautotrophic   Lithoautotrophic microbes obtain their energy from inorganic sources, such as rocks.

malaria   A disease caused by an *parasite* that is transmitted via the "bite" (actually, through a piercing proboscis) of the *Anopheles* mosquito. Humans who have contracted malaria experience severe chills and fever.

marker gene   A *gene* with a known location and identifiable characteristics. Marker genes are used as landmarks in the study of other genes.

master control gene   A *gene* that originated early in *evolution* and now is present in highly conserved (nearly identical) form in many different organisms, and that gives rise to similar *morphology* or function in the different organisms. For example, some scientists claim that a master control gene is responsible for the development of eyes in *species* as diverse as fruit flies and mice.

megabase   One megabase is equal to one million *base* pairs.

meristem   The site of cell division in plants. Meristematic tissues differentiate into the different cells of a plant. The meristematic tissues are located at plant roots and shoot tips.

microbiology   The scientific study of microscopic organisms. These organisms include *bacteria*, *fungi*, *protozoans*, and *viruses*.

mitochondrial DNA   The *DNA* in the mitochondria of *eukaryotic* cells. Mitochondria are often described as the powerhouses of the cell, because they produce ATP, or adenosine triphosphate, which is a biological energy source. An evolutionary debate over Asia versus Africa as the place of origin of humans has led to discussions of Mitochondrial Eve, the "mother" of the human race. The terminology arises because mitochondrial DNA is passed only from mother to offspring.

molecular biology   The scientific study of molecular structures and activities, including *gene* functions, that are involved in biological processes.

mollusc (mollusk)   Member of a group (phylum) of mostly aquatic invertebrates, such as snails, octopi, squid, and slugs. Most have shells, some of which are modified as internal skeletons.

monocot   A flowering plant with one cotyledon, or initial leaf. Grasses are included among the monocots. (See *dicot*.)

morphology   The outward appearance, form, and structure of an organism or part of an organism, or the study of that appearance, form, and structure.

mutation   An alteration in an organism's genetic material. Usually defined as a heritable change, a mutation can have effects that range from fatal to beneficial.

natural history   The scientific study of animal and plant life.

natural selection   The mechanism of *evolution*, as proposed by Charles Darwin. In natural selection, a *species* gradually changes over time when those individuals that carry certain heritable traits reproduce more successfully than other individuals within the species who lack those traits. Offspring carrying those "favored" traits become more and more predominant, and the organism overall evolves to incorporate those traits. (See *Darwin, Charles*.)

neuron   A nerve cell.

nucleic acid   The genetic material of cells, including the *DNA* and *RNA*. Nucleic acids are strings of nucleotides, each of which consists of sugar, a phosphate group, and an organic *base*.

nucleus   The organelle (a defined structure within a cell) in *eukaryotic* cells that contains the *chromosomes*.

paleontology   The scientific study of animal and plant fossils.

parasite   An organism that obtains its food, shelter, or other necessities from another organism, called the host. Parasites can be external (such as ticks) or internal (such as the parasite that causes *malaria* in humans).

parasitology   The scientific study of *parasites*.

pathogen   A disease-causing microorganism.

pathology   The scientific study of disease.

peptide   *Protein*-like string of *amino acids*.

photosynthesis   The transformation of light energy into organic chemicals. Photosynthesis by green plants is the basis for the vast majority of *food chains* on Earth. (See *trophic level*.)

phylogeny   Evolutionary history of organisms or groups of organisms. The phylogeny of groups of organisms is often represented as a tree with each branch dividing into more specific and more evolutionarily advanced taxa.

physical map of the genome   The physical map of the *genome* is a visual depiction of the location of various landmarks, including *genes*, on the *DNA*.

physiology   The scientific study of life functions and processes in organisms.

phytoplankton   The plant portion of *plankton*.

pioneer species   *Species* that take advantage of newly available *habitats*. An example of pioneer species would be those that arrive on an island soon after a volcano has erupted and altered the landscape of that island.

plankton   Organisms that drift in the surface layer of a sea or lake.

plasmid   Circular pieces of bacterial *DNA* that are often used to pick up foreign DNA for use in *genetic engineering.*

population   In ecological terms, a population is a group of individuals from one *species* that live in one place.

population viability analysis (PVA)   An analytical tool used to determine which threatened *species* are the best candidates to receive protective measures. PVA considers such factors as the species' probable fate, population dynamics, and susceptibility to harming influences.

primordial-soup hypothesis   A hypothesis about how life might have begun on Earth. This hypothesis suggests that a combination of gases, other chemicals, and high temperatures combined to present the conditions critical for the formation of simple organic molecules, and eventually the first living cells.

prion   Proteinaceous infectious particles or malformed *protein*s located in the mammalian brain.

prokaryote   A single-celled organism, such as a bacterium, that has no membrane-bound nuclei. (See *eukaryote, archaea*.)

prostaglandin   A substance that stimulates the body in a variety of ways, including the enhancement of hormonal effects.

protein   An assemblage of *amino acid*s that forms the basis of life. Proteins function as *enzymes*, *hormones*, structural elements like hair or nails, blood components, or in a variety of other ways.

protostome   A division in animal classification that refers to a set of embryonic *characters*, especially how the mouth develops. Protostomes include such animals as worms, insects, and *molluscs (mollusks)*.

protozoa   A eukaryotic, single-celled, or acellular organism that may cause disease. (See *eukaryote*.)

radial symmetry   A physical structure for an animal that is arranged symmetrically around one point, rather than having mirror-image right and left sides, as a *bilaterally symmetrical* animal does. Radially symmetrical animals include anemones, jellyfish, starfish, and sea urchins.

radiation (re: adaptive radiation)   The ability of an ancestral group of similar organisms to diverge into new forms and eventually new *species* by taking on new ways of life. Adaptive radiation typically results when organisms are presented with new, unoccupied habitats.

recombinant organism   A term that usually indicates an organism created through *genetic engineering* that contains *DNA* from another organism.

recruitment limitation (re: diversity)   A hypothesis about the cause of diversity in tropical rain forests: According to this hypothesis, many rain forest trees are unable to disperse over large areas. That inability to disperse limits any individual species, or small group of species, from overtaking the forest. Instead, happenstance dictates *species* diversity.

RNA   Ribonucleic acid. Organisms use different types of RNA in *protein* synthesis. RNA, instead of *DNA*, is also the genetic material of many *virus*es.

selective breeding   A method used to choose and enhance an organism's desired traits. Selective breeding involves mating together individuals that carry desirable, heritable traits in the hope of causing them to pass down those traits to their offspring.

senescing   Aging.

shotgun sequencing   A technique used to sequence the *genome* of an organism. It involves breaking up a *chromosome* into small pieces of different lengths, sequencing the pieces, and then putting the genome back together by using computer software to match overlapping bits of the pieces.

somatic cells   Body cells with two sets of *chromosome*s. Sperm and unfertilized egg cells (called germ cells) have only one set.

species   A naturally interbreeding group of similar organisms. *Populations* that are isolated from each other geographically are still considered to be the same species as long as they naturally interbreed. At the point that the two populations cease to interbreed due to morphological, behavioral, or other changes, the populations become distinct species.

stem cell   Derived from *totipotent* cells of the early embryo, stem cells have the ability to differentiate into essentially any type of human cell.

subspecies   A further delineation of a *species*. Single species are often split into subspecies when the species has two or more distinguishable *morphologies* or behaviors.

sustainable development   A balance between the protection and/or maintenance of an *environment* that supports the diversity of life and the often-deleterious effects of humans on that environment.

systematics   A method of organizing taxa, both *extinct* and *extant*, based on their similarities and differences. (See *taxon*.)

taxon   Any standard grouping of organisms, such as *species*, *genus*, family, order or phylum. The plural of *taxon* is *taxa*.

taxonomy   The scientific study of organism classification.

telomere   The end cap on a *chromosome*.

temperature-dependent sex determination   The condition in which offspring gender is based upon incubation temperature. Temperature-dependent sex determination is seen in turtles, in which higher temperatures generally yield female offspring.

threatened   A classification referring to animals that are at risk of *extinction*.

totipotency   The ability of a cell to differentiate into different types of cells. Totipotency is seen in *embryonic cells*, which differentiate into all of the cell types ultimately present in an adult.

transect   An area of set size that is used by field biologists to sample *species* abundance. The exact location of each transect within a larger area is usually randomly determined so that, when considered collectively, the transects provide a statistically relevant sample of the area under study.

transgenic animal   Genetically engineered animal. (See *genetic engineering*.)

transgenic crop   Agricultural crops that have been genetically engineered. (See *genetic engineering*.)

transposable element   A genetic element that can become relocated to another chromosomal location.

tree of life   A depiction of the *evolution* of living things into a tree-like pattern with *branches* representing different groups. Trees may also incorporate a time element to show when groups appeared or divided.

troglomorphy   The set of morphological conditions common to cave-dwelling animals. Reduced or absent eyes is an example.

trophic level   Often described as "a step in the *food chain*." For example, food chain diagrams often place green plants at the base and herbivores, which feed on the plants, at the next-highest trophic level. Food chains usually consist of only three or four trophic levels.

ultraviolet (UV) radiation   UV radiation, which is one type of electromagnetic radiation, has very short wavelengths—150–4,000 angstrom units (Å). UV radiation is divided into three types: UV-A, *UV-B,* and UV-C. Only UV-A and UV-B, which have shorter wavelengths than UV-C, reach Earth's surface. Both UV-A and UV-B can cause health problems for humans, including cancer and cataracts.

UV-B radiation   A type of *ultraviolet radiation*. Scientists have become increasingly interested in the higher levels of UV-B that are reaching Earth due to depletion of the atmosphere's protective ozone layer, which absorbs much of the UV-B before it reaches the planet's surface. With ozone depletion, UV-B levels on Earth have risen. Many biologists are studying the effects of that increase, and some believe it is at least partly responsible for diminished reproductive success among amphibians.

viable   Fertilized eggs are viable if they produce young capable of surviving into adulthood.

virus   An acellular "organism" made up of *protein*-surrounded *RNA* or *DNA*, and little or nothing else. Replication occurs buy infecting a host and injecting the viral *nucleic acid* into a host cell, where the virus particles multiply.

xenotransplantation   The transplantation of organs into humans from other animals, such as pigs. Xenotransplantation has been largely unsuccessful due to human rejection of alien organs, along with concerns about the possibility of cross-species disease transferrals.

yeast artificial chromosome (YAC)   A constructed *chromosome* that scientists use to clone pieces of *DNA*. The YAC serves as a vector for foreign DNA to enter *host cells* and to replicate.

zoology   The scientific study of animals.

INDEX

by Virgil Diodato

American Association of Anatomists
(AAA), 231
American Biology Teacher, The, 227
American Ecological Society, 125
American Entomologist, 239
American Fisheries Society (AFS), 231–
32
American Institute of Biological
Sciences (AIBS), 117, 224–25
American Journal of Botany, 237
American Midland Naturalist, 120
American Museum of Natural History,
64, 130
American Naturalist, 119, 234–35
American Ornithologists' Union (AOU),
232
American redstarts, 29
American Scientist, 229
American Society for Biochemistry and
Molecular Biology (ASBMB), 232
American Society for Bioethics and
Humanities (ASBH), 225
American Society for Horticultural
Science (ASHS), 232–33
American Society for Microbiology
(ASM), 122, 233
American Society of Cell Biology, 126
American Society of Ichthyologists and
Herpetologists (ASIH), 131, 233–
34
American Society of Limnology and
Oceanography (ASLO), 119, 234
American Society of Naturalists, 234–
35
American Society of Plant Physiologists
(ASPP), 235
American Society of Plant Taxonomists
(ASPT), 117, 235
American Zoologist, 242
Ames Research Center, 18
Amino acids, 55, 115
definition of, 245
Amino groups, 115
definition of, 245
Amphibians. *See also specific amphib-
ians*
behavior patterns in, 36
as bio-indicators, 22–23
malformations in, 26, 130–31, 210
medicines from, 20

organizations for study of, 233–34,
243
population declines in, 22–26, 36,
42, 238
Amur tigers, 48
Anaerobic organisms, 15
definition of, 245
Anapsids, 68
definition of, 245
Anatomical Record, The, 231
Anatomists, 231
Anderson, Brian G., 63
Anderson, W. French, 102
Angiosperms (flowering plants), 58, 117
definition of, 245
Animal behavior, 36–37, 67
organizations for, 235–37
Animal Behavior Society (ABS), 235–36
Animal Behaviour, 236
Animal Breeding Research Organiza-
tion, 132
Animal model, for drug development,
54
Animals. *See also specific species*
cloning of, 83–86
development of, 93–94, 126–27,
130–31. *See also* Genetics
new, 11
Annals of the ESA, 239
Annelid, definition of, 245
Anopheles mosquito, 251
Anthropological genetics
definition of, 245
human family tree and, 70
Antibiotics
definition of, 246
resistance to, 54, 77, 89–91
Antibodies, 80, 123
Ants, 37, 218
AOU. *See* American Ornithologists'
Union
Aphids, 106
Apligraf, 81
Appendages
development of, 93, 126–27
malformations in, 26, 130–31, 210
*Aquatic Ecosystem Health and Manage-
ment,* 236
Aquatic Ecosystem Health and Manage-
ment Society, 236

ABOUT THE AUTHOR

Leslie A. Mertz, Ph.D., has been writing about biology, medicine, and other sciences for nearly two decades. A field biologist who has conducted research in Michigan, Bermuda, and the Canary Islands, Mertz currently is teaching biology, writing, and continuing her studies of the natural world.